T0340220

Intermodal Freight Transport and Logistics

Intermodal Freight Transport and Logistics

Edited by
Jason Monios and Rickard Bergqvist

CRC Press is an imprint of the
Taylor & Francis Group, an **informa** business

CRC Press
Taylor & Francis Group
6000 Broken Sound Parkway NW, Suite 300
Boca Raton, FL 33487-2742

First issued in paperback 2019

ISBN-13: 978-1-4987-8512-9 (hbk)
ISBN-13: 978-0-367-89029-2 (pbk)

Visit the Taylor & Francis Web site at
http://www.taylorandfrancis.com

and the CRC Press Web site at
http://www.crcpress.com

Contents

Figures

Tables

Preface

The motivation for producing this book was that, while there have been many academic publications on intermodal transport over the last decade, there is no currently available textbook suitable for the growing number of (particularly postgraduate) students studying this topic, such as our own students at the University of Gothenburg. There are some older books, such as Harris and Schmid (2003) and Lowe (2005), but, as well as providing an up-to-date volume, this book aims to link the operational aspects with the academic approaches published in recent years.

This book is a textbook for students rather than a monograph or research-based book. The goal is to describe how the system works, including a strong focus on practical operational aspects, and then to provide students with a number of frameworks that give context to the system, and finally to demonstrate tools for the analysis of some of the challenges typically associated with intermodal transport systems. More information on the practical aspects of intermodal systems is available online, and students are encouraged to search online for images of different kinds of equipment and videos of operations, such as loading and unloading.

We would like to take this opportunity to thank all the authors who contributed their knowledge and expertise to this book, as well as all those colleagues involved in empirical research on intermodal transport whose work we have drawn upon in the writing of this book.

Jason Monios and Rickard Bergqvist
Gothenburg, Sweden

References

Harris, N. G., Schmid, F. (2003). *Planning Freight Railways.* London: A&N Harris.
Lowe, D. (2005). *Intermodal Freight Transport.* Oxford: Elsevier Butterworth-Heinemann.

Preface

References

Editors

Jason Monios is Associate Professor in Transport Planning and Geography and head of the Freight Transport and Logistics Group at the Transport Research Institute, Edinburgh Napier University, UK. His primary research areas are intermodal transport planning and the geography of port systems, with a specific interest in how these two subjects intersect in the port hinterland. Jason has over 50 peer-reviewed academic publications in addition to numerous research and consultancy reports, covering Europe, North and South America, Asia, the Middle East and Africa. He has co-authored technical reports with UNCTAD and UNECLAC and been adviser to the Scottish Parliament. Recent book publications include *Institutional Challenges to Intermodal Transport and Logistics* (2014) and *Intermodal Freight Terminals: A Life Cycle Governance Framework* (2016). Jason is currently a visiting researcher at the University of Gothenburg, Sweden.

Rickard Bergqvist is Professor in Logistics and Transport Economics at the School of Business, Economics and Law at the University of Gothenburg. His key research areas are maritime logistics, regional logistics, intermodal transportation, dry ports and public–private collaboration. His major works include over 40 refereed journal articles, conference papers and book chapters related to intermodal transport, dry ports, economic modelling, maritime economics and public–private collaboration.

Their recent co-authored book is *Intermodal Freight Terminals: A Life Cycle Governance Framework* (Routledge, 2016).

Editors

Contributors

Abhinayan Basu Bal is assistant professor at the School of Business, Economics and Law, University of Gothenburg. He lectures and conducts research in the field of international trade and maritime law. His research interests lie in maritime law, paperless trade and trade finance. Funding from the European Union, national governments and industry bodies has supported some of his research endeavours. Since 2007, he has actively participated in the deliberations of the United Nations Commission on International Trade Law (UNCITRAL) Working Group III (Transport Law) and Working Group IV (Electronic Commerce). He has published monographs, chapters in books and articles in peer-reviewed journals. Earlier, he worked at Lund University and was responsible for the delivery of the master's programme in Maritime Law. He read for his PhD degree at the World Maritime University, Sweden. He received his LLM in Maritime Law from University College London, UK and holds LLB and MBL degrees from India.

Sönke Behrends is assistant professor at the division of Service Management and Logistics at Chalmers University of Technology, Sweden, and the manager of the Urban Freight Platform (UFP) based in Gothenburg, Sweden. His doctoral thesis 'Urban freight transport sustainability – The interaction of urban freight and intermodal transport', presented in 2012, explored the integration of urban freight planning and intermodal transport networks. Sönke's primary research interest is to advance the understanding of the interaction between logistics, transport and land use and to develop solutions to improve the sustainability of urban logistics operations. His expertise is in the impact assessment of logistics measures, urban freight planning and city logistics. Current projects involve the integration of logistics in urban planning, off-peak distribution and freight in public transport lanes.

Jürgen Wilhelm Böse attended the Braunschweig University of Technology in Germany, where he studied electrical engineering and economics. In 1997, he graduated with a diploma of industrial engineering, specialising in transport information systems. Subsequently, he was a PhD student at the Chair of Information Systems at the same university, and completed his doctorate on planning instruments for process innovations. From 2003 to 2005, he was the head of office for the Centre of Excellence for Traffic and Logistics at the Osnabrück University of Applied Science, Germany. Subsequently, he worked for the international consulting firm HPC Hamburg Port Consulting and carried out numerous projects on container terminal planning. Since 2010, Dr. J.W. Böse has been the chief engineer for the Institute of Maritime Logistics at the Hamburg University of Technology, Germany. His research focuses primarily on the global network of container transport services which has resulted in several projects and publications.

Teodor Gabriel Crainic is professor of operations research, logistics and transportation and logistics management chair, School of Management, Université du Québec à Montréal, Canada. He is also adjunct professor with the Department of Computer Science and Operations Research of the Université de Montréal, senior scientist at the Interuniversity Research Centre for Enterprise Networks, Logistics and Transportation (CIRRELT) and director of its Intelligent Transportation Systems Laboratory. Professor Crainic is associate editor for *Transportation Science*, area editor for the *Journal of Heuristics* and serves on several other editorial boards. He was president of the Transportation Science and Logistics Society of the Institute for Operations Research and the Management Sciences (INFORMS), received the Merit Award of the Canadian Operational Research Society and is a member of the Royal Society of Canada. His research interests are in network, integer and combinatorial optimisation, meta-heuristics and parallel computing applied to the planning and management of complex systems, particularly in transportation and logistics.

Jonas Flodén has a PhD in business administration focusing on logistics. He is a senior lecturer at the Department of Business Administration, School of Business, Economics and Law at the University of Gothenburg, Sweden. Dr. Flodén received his PhD in 2007 on a thesis on modelling of intermodal transport systems. His main research areas are intermodal freight transport and business models. His teaching areas include intermodal transport, supply chain management and information systems.

Erik Fridell obtained a PhD in physics in 1993 and is assistant director and team leader for the Emission and Transport Group at IVL Swedish Environmental Research Institute and adjunct professor in maritime environment at Chalmers University of Technology. Sweden. He has much experience in research on traffic emissions to air, including emission modelling and research on emission abatement strategies. Recent research interests include scenario development, policy instruments and assessing the impact on the environment and health with a focus on new fuels and shipping. He has wide experience in the development and application of various methodologies for calculating and streamlining fuel consumption and air emissions for the transport sector.

Mike Hewitt is associate professor in the Information Systems and Supply Chain Management Department in the Quinlan School of Business at Loyola University Chicago. His research includes developing quantitative models of decisions found in the transportation and supply chain management domains, particularly in freight transportation and home delivery. His work has assisted in the decision making of companies such as Exxon Mobil, Saia Motor Freight and Yellow Roadway. He has expanded his area of expertise to include workforce planning, including working on multidisciplinary projects at the intersection of operations management and cognitive psychology. His research has been funded by agencies such as the National Science Foundation, the Material Handling Institute and the New York State Health Foundation.

Cathy Macharis is professor at the Vrije Universiteit Brussel in Belgium. She is head of the Mobility, Logistics and Automotive Technology Research Centre (MOBI). She teaches courses in operations and logistics management as well as in transport and sustainable mobility. She has been involved in several national and European research projects dealing with topics such as the location of intermodal terminals, the assessment of policy measures in the field of logistics and sustainable mobility, electric and hybrid vehicles. She is chairwoman of the Brussels Mobility Commission.

Dries Meers holds a BSc and an MSc in geography from the KU Leuven (the University of Leuven), Belgium and a PhD in business economics from the Vrije Universiteit Brussel. Since 2011, he has worked as researcher in the MOBI research centre, where he completed his PhD titled, 'From mental to modal shift: Decision support for intermodal transport' in 2016. His research focuses on the use of decision support models to stimulate intermodal freight transport. Central to his research is the further development of the location analysis model for Belgian intermodal terminals (LAMBIT). By applying multi-criteria analysis and discrete choice experiments, he aims to better simulate modal choice behaviour. He was involved in different national research projects dealing with intermodal transport, location analysis and decision support systems.

Ron van Duin is applied research professor of port and city logistics at the Rotterdam University of Applied Sciences, the Netherlands and assistant professor at the Department of Engineering Systems and Services at Delft University of Technology, the Netherlands. He completed his undergraduate studies in econometrics at the Erasmus University Rotterdam (1988). He received his doctorate in technology, policy and management from Delft University of Technology (2012). As a researcher, he has worked on numerous studies concerning (city) logistics, (intermodal) freight transport, infrastructure, ports and terminals. His main interests are in research on sustainability, efficiencies, cost and quality impacts of new technologies in freight transport and logistics.

Bart Wiegmans is senior researcher and assistant professor at the Department of Transport and Planning, Freight Transport and Transport Networks Group at Delft University of Technology. He completed his undergraduate studies in regional and transport economics at the Vrije Universiteit Amsterdam, the Netherlands (1996). He received his doctorate in transport economics from the Vrije Universiteit Amsterdam (2003). As a researcher, he has worked on numerous studies concerning (intermodal) freight transport, infrastructure, ports and terminals. His main interests are in research on efficiencies, cost and quality impacts of new technologies in intermodal freight transport and infrastructure.

Allan Woodburn is a principal lecturer in freight and logistics at the University of Westminster, London. He is responsible for the MSc Logistics and Supply Chain Management course and is involved in a wide range of teaching, research and consultancy activities in the field of freight transport, both within the UK and internationally. In 2000, Allan completed his doctorate examining the role for rail freight within the supply chain. Since then, his main research focus has been on rail freight policy, planning and operations, particularly investigating issues related to efficiency and sustainability. He has also analysed factors that influence freight mode choice decision making and has carried out assessments of various rail freight markets, notably those involving the use of intermodal technologies. Allan has published widely on the subject of rail freight, including 12 peer-reviewed journal papers, several book chapters and a number of research/consultancy reports.

Johan Woxenius is professor of maritime transport management and logistics at the University of Gothenburg since 2008. He has an MSc (industrial engineering) and a PhD from Chalmers University of Technology. His main research field is maritime and intermodal freight transport and the research covers sustainability, industrial organisation, production and information systems. Increasingly, he has engaged in urban freight research. He is a member of the board of the School of Business, Economics and Law at the University of Gothenburg and leads the university's part in maritime research and the education programme of Lighthouse, the logistics research centre Northern LEAD and the Area of Advance Transport together with Chalmers University of Technology. Johan Woxenius is a member of the Royal Swedish Academy of Engineering Sciences, on the scientific advisory boards of research funding organisations and editorial boards of journals. He is also a member of the Scientific Committee of the World Conference on Transport Research Society.

Part I

INTRODUCTION

Chapter 1

Introduction

Jason Monios and Rickard Bergqvist

INTRODUCTION

This chapter establishes the requirement for this textbook and introduces the topic of intermodal freight transport and logistics, identifying its role within the broader area of freight transport, in addition to its place within logistics. The chapter provides a brief outline of policy issues as well as spatial descriptions of intermodal transport via corridors and nodes before moving on to the operational focus of this book. The chapter provides an overarching view of the key elements of intermodal transport operations and how they relate to questions of policy and planning. The structure of the book (operations, frameworks, analysis) is introduced and a brief outline of each chapter is presented.

THE ORIGINS OF INTERMODAL TRANSPORT

Multimodal transport refers to the use of more than one mode in a transport chain (e.g. road and water), while intermodal refers specifically to a transport movement in which the goods remain within the same loading unit. While wooden boxes had been utilised since the early days of rail, it was not until strong metal containers were developed that true intermodal transport emerged. The efficiencies and hence cost reductions of eliminating excessive handling by keeping the goods within the same unit were demonstrated in the first trials of a container vessel by Malcom McLean in 1956.* The initial container revolution, therefore, took place in ports, as the stevedoring industry was transformed over a few decades from a labour-intensive operation to an increasingly automated activity. Vessels once spent weeks in port being unloaded manually by teams of workers; they can now be discharged of thousands of containers in a matter of hours by large cranes, with the boxes being repositioned in the stacks by automatic guided vehicles. This in turn means that ships can spend a much higher proportion of their time at sea, becoming far more profitable.

* See *The Box* (Levinson, 2006) for a historical account of the advent of containerisation.

As shipping and port operations were transformed by the container revolution, a wave of consolidation and globalisation took place. Shipping lines grew and then merged to form massive companies that spanned the globe. Container ports expanded out of origins as general cargo ports or were built entirely from scratch. Some existing major ports today show their legacy as river ports and require dredging to keep pace with larger vessels with deep drafts (e.g. Hamburg), whereas newer container ports are built in deep water, requiring not dredging but filling in to create the terminal land area (e.g. Maasvlakte 2, Rotterdam). The move to purpose-built facilities with deeper water severed the link between port and city, with job numbers reduced and those remaining moved far from the local community, altering the economic geography of port cities (Hesse, 2013; Martin, 2013).

Most of the new generation of container ports are operated by one of a handful of globalised port terminal operators, such as Hutchison Port Holdings or APM. This is the result of the trend towards consolidation across the industry in the decade leading up to the onset of the global economic crisis in 2008, in which many mergers and acquisitions took place in both shipping liner services and port terminal operations. There has also been much vertical integration, with shipping lines investing in port terminals (e.g. Maersk/APM and others). The increasing integration between shipping lines and ports has created an almost entirely vertically integrated system, from port to shipping line to port, within the same company. The inland part of the chain is the new battleground, but it is more complex than the sea leg.

As a result of these changing industry dynamics, ports changed from city-based centres of local trade to major hubs for cargo to pass through, with distant origins and destinations. This development was driven to a large degree by the container revolution, as distribution centres (DCs) located in key inland locations became key traffic generators. Port hinterlands began to overlap as any port could service the same hinterland. Shipping services were rationalised, with large vessels traversing major routes between a limited number of hub ports. Cargo was then sent inland or feedered to smaller ports. The introduction of new, larger vessels on mainline routes is initiating a process whereby vessels cascade down to other trades. The result is that feeder vessels are being scrapped at a higher rate than normal, and the order book for new builds is at a historic low.

Despite such consolidation in shipping and port terminals, the business of maritime transport remains highly volatile, not just cyclical but dramatically so, exhibiting widely divergent peaks and troughs. According to a senior executive from Maersk, '2009 for Maersk Line was the worst year we have ever had and 2010 was the best – that is not very healthy' (Port Strategy, 2011). Therefore, port actors seek stability where possible, needing to anchor or capture traffic to make themselves less susceptible to revenue loss when the market is low. Inland transport is now the area where they seek to secure this advantage. This need to control inland connections is not just about physical infrastructure but institutional issues such as labour relations and other regulatory issues.

While the rise of intermodal transport originally related primarily to sea transport, the land leg was undertaken by all modes, which were also busy transporting domestic traffic. Before the nineteenth century, hinterland transport primarily consisted of sailing ships and horse-drawn wagons. During the nineteenth century, barge canal operations combined with horse or rail became more common, and there were even some early experiences with intermodal transport units (ITU). One of the first experiences of ITU was in England, where it was used for the transport of coke between road carts, barges and railcars. By the early twentieth century, rail wagons were put on seagoing vessels and trucks on rail wagons. Intermodal transport began, but there were still few systems that could carry a standardised load unit suitable for intermodal transport.

By the mid-twentieth century, the carrying of road vehicles by railcar, known as piggyback transport or trailers-on-flatcars (TOFC), became more widespread. This method of transport was previously introduced in 1822 in Germany, and in 1884 the Long Island Railroad started a service of farm wagons from Long Island to New York City (APL, 2011). As TOFC caught on during the 1950s, the use of boxcars declined. One reason why TOFC became popular was the improved efficiency of cargo handling and the end of break bulk handling. Between 1957 and 1992, the number of boxcars in the United States decreased from about 750,000 to fewer than 200,000 (APL, 2011).

The increasing integration of international and domestic transport was a result of globalised supply chains growing out of relaxed tariff and trade barriers as well as ever-cheaper sea transport. Inputs to manufacturers and even finished products were being imported at a growing rate from cheaper supply locations and, to overcome congestion and administrative delays at ports, shipping lines began to offer inland customs clearance. The bill of lading could now specify an inland origin and destination. These sites were variously known as inland clearance depots (ICDs) and 'dry ports'. In the United Kingdom, so-called 'container bases' were established at key locations around the country in the late 1960s to handle containerised trade passing through south-eastern ports to and from inland locations in the north and centre of the country. These freight facilities were usually located next to intermodal terminals, but the actual transport mode could be road or rail (see Hayuth, 1981; Garnwa et al., 2009). This kind of trade was especially promoted for landlocked countries lacking their own ports. Thus the 'dry port' could offer a gateway role and reduce transport and administrative costs (Beresford and Dubey, 1991).

It was in the United States where true intermodal transport was established successfully. During the 1980s, carriers operating in the trans-Pacific trade were suffering from excess tonnage and low freight rates. To increase its cargo volumes, American President Lines (APL) formed the first transcontinental double-stack rail services, recognising that intermodal transport provided a 10-day time saving compared to the sea route through the Panama Canal to New York (Slack, 1990). While the transit time was important, APL also offered more services to the shipper, as the customer could receive a single through bill of

Table 1.1 Freight transport in different regions (billion tkm)

Region	*EU-28*	*USA*	*Japan*	*China*	*Russia*
Year	*2012*	*2011*	*2012*	*2012*	*2012*
Road	1692.6	2038.9	210.0	5953.5	249.0
Rail	407.2	2649.2[a]	20.5	2918.7	2222.0
Inland waterways	150.0	464.7		2829.6	61.0
Oil pipeline	114.8	968.6		317.7[b]	2453.0
Sea (domestic/intra EU28)	1401.0	263.1	177.6	5341.2	45.0

Source: Authors, based on European Commission, *EU Transport in Figures: Statistical Pocketbook 2014*, Publications Office of the European Union, Luxembourg, 2014.

a Class I rail.

b Oil and gas pipelines.

lading. The growth of discretionary cargo allowed APL and other shipping lines to expand their capacity in the trans-Pacific. By using larger, faster ships, a carrier could offer a fixed, weekly sailing schedule, while the additional capacity reduced per-unit costs. In Europe, intermodal freight transport developed in the 1990s, although the fragmented geographical and operational setting (e.g. national jurisdictions and constraints on interoperability) as well as physical constraints (e.g. limited opportunities for double-stack operation) meant that progress was not as swift nor as successful as in the United States (Charlier and Ridolfi, 1994).

Different geographical regions have substantially different prerequisites for the respective mode of transport. There are, therefore, substantial differences between regions and countries when it comes to the usage of the different modes of transport (Table 1.1). Some of the differences can be explained by geographical conditions, but other important facts are regulatory aspects, status of infrastructure and, occasionally, technology.

From a transport work (tkm) perspective, EU-28 uses road transport extensively. Japan has a similar situation, but Japan's geographical conditions make it more reliant on road transportation than the EU-28. The use of the double-stacking of containers, and hence more loading capacity, is one reason why the United States has a larger share of rail transport compared with the EU-28. Various types of electrical systems, signalling systems, and so on in the European Union (EU) are other reasons why rail has a lower market share in the EU compared with other regions.

SPATIAL CONCEPTS IN INTERMODAL TRANSPORT

While this book focuses mostly on the practical aspects of intermodal transport rather than abstract approaches, a brief overview of the geography of intermodal transport can be helpful for understanding and analysing the intermodal system.

A node may be defined simply as a location or a point in space; in the case of transport, this would represent an origin or destination point. In practice, only nodes of a certain size are considered: where a certain level of units is concentrated, moves through or otherwise utilises this access point. A node can serve as an access point to join a transport network, or it may be a point joining two linkages within a system. Two defining characteristics of such nodes are centrality and intermediacy. A central location exerts a centripetal pull on the region, while intermediacy refers to an intermediate location in between centres. Fleming and Hayuth (1994) observed how central locations are often also intermediate, acting as gateways to other locations. Nodes can also be defined as points of articulation or interfaces between spatial systems (Rodrigue, 2004), particularly different levels (e.g. local and regional) and types (e.g. intermodal connections), but the articulation concept can also include joining different categories of system, such as transport and logistics systems. This involves the relation of the transport activity to other related activities, such as processing, distribution and all activities within the wider logistics system (Hesse and Rodrigue, 2004).

Locations, points or nodes are joined by links. These links may be physical, meaning either fixed, such as roads, rail track and canals, or flexible links such as sea routes. They may also be operational links, referring to services, such as road haulage or shipping schedules. In operational terms, links can be measured in terms of their capacity, current usage and congestion. Nodes are often rated by their connectivity, which could either refer to the number and quality of physical links or the number and frequency of operational links, all of which derive to a certain extent from the qualities of centrality and intermediacy already discussed. Operational strategies of freight operators go beyond single links and can be expressed in various ways, such as hub-and-spoke, string or point-to-point. These combined operational plans then become transport networks, either a single company network or the accumulation of all available services within a given area.

A network can be defined as the set of links between nodes. Again, this may be considered from a physical or operational perspective. The previously discussed issue of connectivity can be used to assess the quality of a node, but it can also refer to the quality of a network in which a number of nodes are connected. A high-quality network may contain a number of nodes with high connectivity, high centrality and high intermediacy, linked to each other with frequent, high capacity services within a small number of degrees.

A corridor can be defined as an accumulation of flows and infrastructure (Rodrigue, 2004). In some ways, the concept of a corridor is somewhat arbitrary and may be used for branding or public relations (PR) purposes. This is because, beyond a specific piece of infrastructure (e.g. one road or rail line between two places), a corridor usually denotes a large swathe of land through which multiple routes are possible along numerous separate pieces of infrastructure with many different flows organised and executed by different actors. The corridor branding concept can be useful for attracting funding and focusing attention on a specific region.

Scholars often prefer a network focus to a corridor focus because it gives a better measure of the wider distribution requirements of each node, whereas a corridor focus can be limiting, given the globalisation of today's distribution patterns. Yet, a corridor approach is more amenable to public planners, who need to coordinate many divergent demands for transport and land use, within local, regional, national and international policy, planning and funding regimes.

Classification of inland freight facilities and the activities in which they engage is difficult, and, despite some earlier analysis of their functions and locations (Hayuth, 1980; Slack, 1990; Wiegmans et al., 1999), it is only in recent years that their spatial and institutional characteristics have begun to be treated in detail. Firm definitions are difficult as each site is different; therefore, it is best to focus on the key aspects of each. Rodrigue et al. (2010) related the multiplicity of terms to the variety of geographical settings, functions, regulatory settings and the related range of relevant actors, and proposed that the key distinction is between transport functions (e.g. transloading between modes, satellite overspill terminals or load centres) and supply chain functions (e.g. storage, processing, value-added). Table 1.2 lists the main inland terminal classifications used in the literature.

It can be seen from Table 1.2 that, as Rodrigue et al. (2010) argue, inland freight nodes can be divided into two key aspects: the transport function and the supply chain function, with each classification exhibiting various aspects of each. Categories 5, 7 and 8 in Table 1.2 specifically require an intermodal transport connection, while all the others relate to other functions, such as customs, warehousing, consolidation, logistics and other supply chain activities. In practice, many of these sites would have intermodal connections, but it is not specifically required within the categorisation.

POLICY AND PLANNING FOR INTERMODAL TRANSPORT

It is a public responsibility to ensure sufficient capacity on all transport links to support a growing economy, but the mix of public and private interests in freight operations can result in considerable uncertainty when it comes to investment in upgrades and capacity enhancements, or connecting freight nodes to the transport network. While highways and motorways are generally maintained by governments for both passenger and freight use, rail and waterways can be either privately or publicly owned. Interchange sites such as ports and rail/barge terminals may exhibit a variety of ownership, management and governance regimes (see Chapter 8). Where they are under public control, rail and waterways are for the most part simply maintained in their current state, with the occasional new section or upgrade, but the high levels of public investment expended for the apparent benefit of private companies can be contentious.

Table 1.2 Inland freight node taxonomies

No.	*Name*	*Description*
1	Inland clearance (or container) depot	The focus here is on the ability to clear customs at the inland origin/destination site rather than at the port. Started to spring up in the 1960s. A kind of warehouse area (could just be small) with customs clearance. Any transport mode is acceptable within this definition. See Hayuth (1980), Beresford and Dubey (1991), Garnwa et al. (2009), Pettit and Beresford (2009).
2	Container freight station	This is basically a shed for container stuffing/stripping/(de-) consolidation. It is not a node in itself, but more a service that may be provided within a port or an inland site.
3	Dry port I	Synonymous with ICD, either in a landlocked country or one that has its own seaports (see Beresford and Dubey, 1991; Garnwa et al., 2009).
4	Inland port	Favoured in the United States (see Rodrigue et al., 2010). Customs is less of an issue in the United States because 89% of its freight is domestic. As the railroads run on their own private track, terminals are also private nodes, so the management of containers is a closed system for that firm to manage the flow. Some reservations to using it in Europe because there an inland port generally has water access, and inland terminals are not normally the massive gateway nodes that they are in the United States (i.e. fewer than 100,000 lifts annually vs. many times that in the United States).
5	Intermodal terminal	Generic term for an intermodal interchange, that is, road/rail, road/barge. Could strictly speaking be just the terminal with no services or storage nearby, but would generally involve such services. It is also referred to as transmodal centre by Rodrigue et al. (2010), which draws attention to its primary function, interchange rather than servicing an O/D market. In practice, presumably do some O/D freight would be serviced as well to make the site more feasible.
6	Freight village, logistics platform, interporto, GVZ, ZAL, distripark (if located in or near a port)	These are big sites with many sheds for warehousing, logistics, and so on, and usually relevant services too. May have intermodal terminal or may be road only. May have customs or may not. Distripark is used to denote a site based within or on the outskirts of a port. See Notteboom and Rodrigue (2009), Pettit and Beresford (2009).
7	Extended gate	Specific kind of intermodal service whereby the port and the inland node are operated by the same operator, managing container flows within a closed system, thus achieving greater efficiency. The shipper can leave or pick up the container at the inland node just as with a port. See Van Klink (2000), Rodrigue and Notteboom (2009), Roso et al. (2009), Veenstra et al. (2012).

(Continued)

Table 1.2 (Continued) Inland freight node taxonomies

No.	*Name*	*Description*
8	Dry port 2	New definition by Roso et al. (2009). This would seem to be an ICD with large logistics area and intermodal (rail or barge) connection to the port, in combination with extended gate functionality, thus providing an integrated intermodal container handling service between the port and the fully serviced inland node.
9	Satellite terminal	See Slack (1999). Usually a close-distance overspill site, operated almost as if it is part of the port. Could be considered a short-distance extended gate concept. This should really be rail-connected but some sites are linked by road shuttles. (That would seem to ignore the main function, which is to overcome congestion, but it can reduce congestion by reducing the time each truck spends in the port on administrative matters.)
10	Load centre	This concept could apply to inland terminals or ports, but in the case of the former it refers to a large inland terminal to service a large region of production or consumption. Probably the classic kind of inland node as it serves as a gateway to a large region. Tends to fit well with the American inland port typology. It normally refers specifically to the terminal, but generally in this sort of location one would expect to have a lot of warehousing, and so on in the area if not part of the actual site. See Slack (1990), Notteboom and Rodrigue (2005), Rodrigue and Notteboom (2009).

Source: Monios, J., *Institutional Challenges to Intermodal Transport and Logistics*, Ashgate, London, 2014.

The success of intermodal transport in the United States is partly a result of a vertically integrated system in which rail operators own and manage both their infrastructure network and the operations. The United States is large enough to sustain competition between different operators, each with their own extensive infrastructure network serving most of the same origins and destinations. A smaller geographical region such as Europe would find such a system difficult. In the current system across Europe, infrastructure is owned by national governments, while individual rail operators compete with one another to run services, paying access charges for their use of the track infrastructure. In the United Kingdom, these companies are private, whereas in Europe they are a mixture of private and public. However, even where they are public, evolving rail reform in Europe due to EU policy has required that they operate as quasi-private companies, with full organisational separation from the infrastructure-owning parts of the same national companies. An interesting comparison is China, which is still publicly controlled and divided into several vertically integrated regions.

From an operational perspective, in terms of the impact of rail infrastructure on successful intermodal operations (see Chapter 3), there are rail gauge (width between the rails) compatibility issues between some countries, such as between continental Europe and the Iberian peninsula; and in the United Kingdom, loading gauge (width and height) is restricted due to bridges and tunnels, meaning that high cube containers cannot be carried on some parts of the network unless low wagons are used, adding expense and inconvenience. The other site of interaction between infrastructure and operation is intermodal terminal development (discussed in Chapter 6).

Until recent times, operational decisions and mode choice were the preserve of the industry. The inland leg was taken by road, rail or inland waterway, according to the economic and practical imperatives of the shipper and transport provider. Rail and water generally dominated long hauls because they were cheaper, whereas road haulage would perform shorter journeys, particularly any journey where its natural flexibility and responsiveness made it the natural choice. The role of the public sector was operational in some countries where rail was nationally owned, but otherwise related mostly to regulation. However, as emissions and congestion rose up the government agenda in the 1990s, governments began to see their role as more directly interventionist in order to address the negative externalities of transport.

The European Union transport policy document published in 2001 (European Commission, 2001) made a clear statement in favour of supporting intermodal transport as one method of reducing emissions and congestion (Lowe, 2005). In the late 1990s and early 2000s, policy documents multiplied across Europe promising support for greener transport measures to reduce dependence on road transport, while also taking care politically not to be seen to threaten the performance of the road haulage industry which remains essential to a functioning transport system. While road haulage produces more emissions than other modes per tonne-kilometre, lorries are increasingly environmentally friendly, due in part to government regulation (Chapter 13). Factors encouraging modal shift away from road haulage include continuing fuel price rises and, in Europe, the working time directive limiting driver hours per week (although there are questions as to how closely such regulations are followed in different parts of Europe, not to mention the lack of such regulation in other parts of the world). Road user charging is another policy implemented in some parts of Europe (e.g. the Maut scheme implemented in Germany which charges lorries for use of the motorways). Better fleet management, use of Information and Communications Technology (ICT), increased backhauling, triangulation, reverse logistics, returning packaging for recycling and other operational measures (McKinnon, 2010; McKinnon and Edwards, 2012), mean that emissions (if not congestion) can be reduced quite substantially through improvements to road operations rather than through modal shift.

The establishment of the Single European Market in 1993 and the increasing integration throughout the EU, including customs union and almost-complete currency union, altered distribution strategies and increased cross-border freight movements, including a change in the location of DCs, as companies developed pan-European rather than national distribution strategies. A series of European directives drove progress towards harmonising administrative and infrastructure interoperability between member states. A cornerstone of these efforts in Europe is the Trans-European Network–Transport (TEN-T) programme, which identifies high-priority transport linkages across Europe; member states can then bid for funding to invest in upgrading these links. It covers both passenger and freight and includes all modes (as well as 'motorways of the sea'). Its primary goal is not modal shift per se but increased connectivity between member states.

Logistics clusters have many agglomeration benefits both for business and for transport. However, while access to a large transport corridor, especially an intermodal corridor, may reduce emissions over the length of the journey, it will increase emissions around the access point where traffic is congregated. Linking a town to a nearby corridor may bring economic benefits through direct and indirect jobs, but it may increase emissions and raise property prices and other aspects explored by economic geographers. Increasingly, transport geography looks to economic geography to investigate how transport policy links with economic development policy. Much European funding for transport projects is aimed at reducing emissions, but it is actually pursued by local and regional bodies because they desire economic benefits from logistics development.

The role of the US Federal Government with regard to transport has been primarily related to safety and licencing regulation, but it is increasingly taking a direct role in intermodal infrastructure and operations, aiming to promote both emissions reduction through modal shift and economic growth through improving access for peripheral regions. For example, the Transportation Investment Generating Economic Recovery (TIGER) programme, as part of the US stimulus package, provided $1.5 billion in federal funding in 2009, to be bid for by consortia of public and private partners across the country (Monios, 2014).

These incentives are less common in developing countries, which are focused more on developing their logistics infrastructure to support business; therefore interventionist transport policy has been pursued primarily in developed countries. However, supranational development agencies, such as the United Nations Conference on Trade and Development (UNCTAD) and the United Nations Economic and Social Commission for Asia and the Pacific (UNESCAP), have promoted policy actions to improve port-hinterland connections and logistics performance in developing countries, especially for landlocked or otherwise poorly-connected inland regions (e.g. UNCTAD, 2004, 2013; UNESCAP, 2006, 2008). Yet it is important to remember that in many cases there are good business reasons why intermodal transport is not flourishing at a certain location due to cargo and route characteristics. Government money is not always the answer unless the industrial organisation can be improved, thus requiring an appreciation of the institutional challenges to intermodal transport and logistics.

The other important role for governments is the regulation, administration and bureaucracy of trade facilitation. Within a country, it will involve licences to operate transport vehicles, regulation of vehicle and infrastructure quality and quantity and permission to operate as a commercial transport provider. It will also cover planning permission to build a logistics platform, provide connections to electricity and water services, incorporating related issues, such as noise for local residents and all the small issues of local planning. At the international level, regulation and international treaties cover the use of bills of lading and various transport and insurance contracts that must be legally approved (Chapter 10), as well as customs legislation in each country.

STRUCTURE OF THE BOOK AND AUTHOR CONTRIBUTIONS

This book is a textbook for students, rather than a monograph or research-based book. The goal is to describe how the system works, including a strong focus on practical operational aspects; to provide students with a number of frameworks that give context to the system; and, finally, to demonstrate some tools for analysis of some of the main problems associated with intermodal transport systems. Much more information on the practical aspects of operations is available online, and students are encouraged to search online for more images of equipment and videos of operations, such as loading and unloading.

The book is structured in three sections: operations, frameworks and analysis. The first section focuses on the practical aspects of intermodal transport, beginning with an overview of the main vehicle and equipment types used in the industry. Then, each of the three inland transport modes are addressed in their own chapters: rail, barge and road. Each of these chapters describes how the system works and presents the main issues arising. Chapter 6 examines the design and operation of the intermodal terminal itself, and Chapter 7 looks at the port interface.

The second section introduces a number of frameworks within which to understand the context of intermodal transport. Chapter 8 examines the intermodal system from a perspective of management and economics, discussing business models for service provision and economic optimisation from a system perspective. Chapter 9 takes a logistics focus, discussing how shippers can modify their supply chain management in order to match with the requirements for intermodal transport. Chapter 10 introduces a legal perspective, covering the issues of contracts and responsibility for the shipment throughout the intermodal transport chain.

The third section introduces three methods of analysis of an intermodal system. Chapter 11 looks at design and modelling of intermodal systems, covering how to transform a real system into a conceptual model and then how to analyse the different elements. Chapter 12 takes an operations research perspective on analysis of intermodal operations. Finally, Chapter 13 covers the environmental aspects of intermodal transport, discussing how to measure and compare different kinds of emissions and various policy perspectives to manage these issues.

REFERENCES

APL. (2011). Evolution of rail in America. Available at http://www.apl.com/history/html/overview_innovate_rail.html. Accessed 21 January 2011.

Beresford, A. K. C., Dubey, R. C. (1991). *Handbook on the Management and Operation of Dry Ports.* RDP/LDC/7. Geneva, Switzerland: UNCTAD.

Charlier, J. J., Ridolfi, G. (1994). Intermodal transportation in Europe: Of modes, corridors and nodes. *Maritime Policy and Management.* 21 (3): 237–250.

European Commission. (2001). *European Transport Policy for 2010: Time to Decide.* Luxembourg: European Commission.

European Commission. (2014). *EU Transport in Figures: Statistical Pocketbook 2014.* Luxembourg: Publications Office of the European Union.

Fleming, D. K., Hayuth, Y. (1994). Spatial characteristics of transportation hubs: Centrality and intermediacy. *Journal of Transport Geography.* 2 (1): 3–18.

Garnwa, P., Beresford, A., Pettit, S. (2009). Dry ports: A comparative study of the United Kingdom and Nigeria. In UNESCAP, *Transport and Communications Bulletin for Asia and the Pacific No. 78: Development of Dry Ports.* New York: UNESCAP.

Hayuth, Y. (1980). Inland container terminal: Function and rationale. *Maritime Policy and Management.* 7 (4): 283–289.

Hayuth, Y. (1981). Containerization and the load center concept. *Economic Geography.* 57 (2): 160–176.

Hesse, M. (2013). Cities and flows: Re-asserting a relationship as fundamental as it is delicate. *Journal of Transport Geography.* 29: 33–42.

Hesse, M., Rodrigue, J.-P. (2004). The transport geography of logistics and freight distribution. *Journal of Transport Geography.* 12 (3): 171–184.

Levinson, M. (2006). *The Box: How the Shipping Container Made the World Smaller and the World Economy Bigger.* Princeton, NJ: Princeton University Press.

Lowe, D. (2005). *Intermodal Freight Transport.* Oxford, UK: Elsevier Butterworth-Heinemann.

Martin, C. (2013). Shipping container mobilities, seamless compatibility and the global surface of logistical integration. *Environment and Planning A.* 45 (5): 1021–1036.

McKinnon, A. (2010). *Britain without Double-Deck Lorries.* Edinburgh, Scotland: Heriot-Watt University.

McKinnon, A., Edwards, J. (2012). Opportunities for improving vehicle utilisation. In McKinnon, A., Browne, M., Whiteing, A. (Eds.), *Green Logistics: Improving the Environmental Sustainability of Logistics*, 2nd Ed., pp. 205–222. London: Kogan Page.

Monios, J. (2014). *Institutional Challenges to Intermodal Transport and Logistics.* London: Ashgate.

Notteboom, T. E., Rodrigue, J. (2005). Port regionalization: Towards a new phase in port development. *Maritime Policy and Management.* 32 (3): 297–313.

Notteboom, T. E., Rodrigue, J.-P. (2009). Inland terminals within North American and European Supply Chains. In UNESCAP, *Transport and Communications Bulletin for Asia and the Pacific No. 78: Development of Dry Ports.* New York: UNESCAP.

Pettit, S. J., Beresford, A. K. C. (2009). Port development: From gateways to logistics hubs. *Maritime Policy and Management.* 36 (3): 253–267.

Port Strategy. (2011). Maersk calls ports to the table. Port Strategy. Mercator Media. 17 October 2011. Available at http://www.portstrategy.com/news101/products-and-services/maersk-calls-ports-to-the-table. Accessed 2 September 2013.

Rodrigue, J.-P. (2004). Freight, gateways and mega-urban regions: The logistical integration of the Bostwash corridor. *Tijdschrift voor Economische en Sociale Geografie*. 95 (2): 147–161.

Rodrigue, J.-P., Debrie, J., Fremont, A., Gouvernal, E. (2010). Functions and actors of inland ports: European and North American dynamics. *Journal of Transport Geography*. 18 (4): 519–529.

Rodrigue, J.-P., Notteboom, T. (2009). The terminalisation of supply chains: Reassessing the role of terminals in port/hinterland logistical relationships. *Maritime Policy and Management*. 36 (2): 165–183.

Roso, V., Woxenius, J., Lumsden, K. (2009). The dry port concept: Connecting container seaports with the hinterland. *Journal of Transport Geography*. 17 (5): 338–345.

Slack, B. (1990). Intermodal transportation in North America and the development of inland load centres. *Professional Geographer*. 42 (1): 72–83.

Slack, B. (1999). Satellite terminals: A local solution to hub congestion? *Journal of Transport Geography*. 7 (4): 241–246.

UNCTAD. (2004). *Assessment of a Seaport Land Interface: An Analytical Framework*. Geneva, Switzerland: UNCTAD.

UNCTAD. (2013). *The Way to the Ocean: Transit Corridors Servicing the Trade of Landlocked Developing Countries*. Geneva, Switzerland: UNCTAD.

UNESCAP. (2006). *Cross-Cutting Issues for Managing Globalization Related to Trade and Transport: Promoting Dry Ports as a Means of Sharing the Benefits of Globalization with Inland Locations*. Bangkok, Thailand: UNESCAP.

UNESCAP. (2008). *Policy Framework for the Development of Intermodal Interfaces as Part of an Integrated Transport Network in Asia*. Bangkok, Thailand: UNESCAP.

Van Klink, H. A. (2000). Optimisation of land access to sea ports. In 'Land Access to Sea Ports,' European Conference of Ministers of Transport, Round table 113, Paris, 10–11 December 1998, pp. 121–141.

Veenstra, A., Zuidwijk, R., Van Asperen, E. (2012). The extended gate concept for container terminals: expanding the notion of dry ports. *Maritime Economics and Logistics*. 14 (1): 14–32.

Wiegmans, B. W., Masurel, E., Nijkamp, P. (1999). Intermodal freight terminals: An analysis of the terminal market. *Transportation Planning and Technology*. 23 (2): 105–128.

Part II

OPERATIONS

Chapter 2

Intermodal transport equipment

Rickard Bergqvist and Jason Monios

INTRODUCTION

This chapter provides a thorough list and description of different types of locomotives, trucks, coastal and inland vessels, cranes, reach stackers and so on, as well as intermodal loading units, wagons, containers, chassis, swap bodies, trailers and so on. Photos and diagrams are used with discussion of dimensions and capacity. In terms of equipment, issues discussed include high cube and pallet-wide containers, container owning and leasing, issues of empty container management, swap bodies versus trailers, different wagon designs and ownership models. More detailed discussion of road, rail and water vehicles and vessels is provided in the relevant mode-specific chapters. It should be noted that this chapter focuses on the most common equipment and does not go into technical detail. Most manufacturers of the equipment mentioned in this chapter have excellent websites with many photos and full technical details, so interested students are encouraged to visit these sites for more information and to explore the wide variety of equipment in addition to that listed in this chapter.

LOADING UNITS

Unitised transport refers to the movement of freight in a standardised loading unit, which may be a single consignment of goods or may be a groupage load of smaller consignments managed by a freight forwarder. The unit in question, often referred to as an intermodal transport unit (ITU) or intermodal loading unit (ILU), may be an International Organization for Standardization (ISO) maritime container, a swap body or a semi-trailer.

Containers

Container types

ISO containers are the strongest loading unit, as well as being stackable. They are, therefore, the most versatile. The key underpinning of successful intermodal transport was not simply the invention or adoption of these containers but their increasing standardisation. This was a long process (see Levinson [2006] for a detailed history and the role of the ISO) that resulted in a handful of main container types. Three important ISO standards regarding freight container specification are

- ISO 668:1995 Freight Container – Classification, dimension and rating
- ISO 6346:1995 Freight Container – Coding, identification and marking
- ISO 1161:1984 Freight Container – Corner fittings – Specification

Today, a wide range of container types are used in international trade. While several lengths, heights and widths still remain, 20 ft and 40 ft long units remain dominant on deep-sea vessels, and containers are therefore measured as multiples of 20 ft (twenty-foot equivalent units or TEU). As this is an approximate measure, the height of the box is not considered. For instance, two containers of 20 ft length, one with height of 9 ft 6 in high cube and the other 4 ft 3 in half height, both count as one TEU. Significant divergence remains, however, particularly domestically. For domestic movements, both the United Kingdom and the United States favour domestic intermodal containers of the same dimensions as road trailers, which means they can carry the same payload (length 45 ft in the United Kingdom and 53 ft in the United States). Standard height is 8 ft 6 in although other heights exist, and 9 ft 6 in (known as high cube) are increasingly common as they allow extra volume, subject to weight limits. Standard width is 8 ft, although again other widths are possible, and in Europe the 8 ft 2 in (known as pallet-wide) is popular because, again, it is closer to the load capacity of a semi-trailer. Refrigerated containers are also wider. The need to accommodate both increased height and width containers has caused many difficulties for rail networks developed in earlier times, such as the United Kingdom's, with loading gauge restrictions caused by bridges and tunnels that require costly upgrade work. In a study of container types moving through U.K. ports in 2010, Monios and Wilmsmeier (2014) found that 72% of containers were 40 ft, 20% were 20 ft and 7% were 45 ft length. The study also found that 59% of 40 ft and 94% of 45 ft were high cube containers, while only a negligible amount of 20 ft were high cube. The vast majority of containers were standard width, but most 45 ft containers were pallet-wide because they are used in intra-European trade, and it is important to achieve the same pallet loading as a truck.

The maximum gross mass for a 20 ft container is 30,480 kg, and for a 40 ft it is 34,000 kg. Allowing for the tare weight* of the container, the maximum payload mass is therefore reduced to approximately 28,380 kg for 20 ft and 30,100 kg for 40 ft containers, meaning that two 20 ft containers can take more weight than one 40 ft container. Twenty-foot containers are favoured for heavier goods because weight limits would mean that the 40 ft container would reach its weight

* Tare weight is the weight of an empty vehicle or container.

limit before filling the space, sometimes referred to in the industry as 'weighting out' before 'cubing out', that is, using up the cubic capacity or volume.

Intra-European shipments move in 45 ft containers, available in both standard height and high cube. In recent years, 'pallet-wide' containers have been developed on European short sea routes. These are 2.4 in wider than standard containers, giving the same internal width as a road trailer (2.44 m), and thus able to fit the same number of pallets. Two different standards of pallet sizes are used. U.K. pallets (GKN or CHEP brands) measure 1200 × 1000 mm, whereas European pallets measure 800 mm × 1200 m.* U.K. pallets are loaded horizontally (2 × 1200 mm) and European Union (EU) pallets vertically (3 × 800 mm). A 45 ft pallet-wide container takes the same number of pallets as a road trailer (26 U.K. or 33 Euro), compared with a 45 ft standard width container (24 U.K. or 27 Euro), a 40 ft pallet-wide container (24 U.K. or 30 Euro) or a 40 ft standard width container (22 U.K. or 25 Euro). There is a move in Europe to make 45 ft pallet-wide maritime containers the industry standard (Bouley, 2012). The problem with this proposal is that most deep-sea ships cannot accommodate these containers in their 20/40 ft cellular holds.

Containers are secured to their transport equipment (e.g. deck of a ship, road trailer or rail wagon) by the corner castings, which are hollow cubes at each corner. The corner casting is placed over the twist lock, which is then turned to secure the fitting (Figure 2.1). Additional containers can then be stacked on top and secured in the same manner.†

General-purpose containers are fully enclosed and usually have a door at one end (although different kinds exist, such as those with side doors, vents, etc. – see Table 2.3 and Figure 2.2). They are the dominant type of container for transporting consumer goods as well as all kinds of dry goods. Generally, the goods will be loaded in boxes, cartons and cases and placed on pallets, but goods can also be loose or in sacks, drums, bales and so on.

Refrigerated containers (known as *reefers*) are equipped with their own cooling systems for transporting temperature-sensitive commodities such as fresh fruit. They provide more flexibility rather than requiring a fully refrigerated hold of the vessel to carry the temperature-sensitive cargo. However, they require constant power; therefore, while in the ship, on the truck/train or on the quay, they need to be plugged into a power source. Some have their own on-board power source that may last several hours for an overland journey but not the days that it may sit in a port or weeks at sea (Figure 2.3).

Open-top containers are similar to general-purpose containers but have no roof. They are used for transporting bulky goods and general cargo that are loaded from the top rather than through the end door. Flat containers have no

* See http://www.searates.com/reference/pallets/ for a useful diagram showing how the two pallet types fit into different container sizes.

† For detailed information see *A Master's Guide to Container Securing* (Murdoch and Tozer, 2012).

Figure 2.1 Close-up of twist-lock on a semi-trailer. (From © selbst [Public domain], via Wikimedia Commons)

Figure 2.2 General-purpose containers stacked at an intermodal terminal. (From Rickard Bergqvist and Jason Monios.)

roof or walls and are used for general cargo and mechanical goods and vehicles, both for reasons of easy loading and also because they may not fit within the profile of the container. Tank containers embed the tank inside a steel frame of standard container dimensions so that they can be handled and transported in the same manner as a regular container. They are used for carrying liquids and gases (Figure 2.4).

Figure 2.3 Temperature-controlled container loaded on a rail wagon. (From Rickard Bergqvist and Jason Monios.)

Some innovative ideas in recent years include foldable containers. Even with the additional handling costs, the large reduction in transport costs means that significant cost savings are possible. The applicability of foldable containers has been studied by Konings (2005a,b) and Shintani et al. (2012). While it has been shown that the concept itself is feasible and could save money, the widespread adoption of these containers by container lessors and shipping lines is required before the value can be exploited by shippers. At present, foldable containers are substantially more expensive than regular containers (about double the cost – prices fluctuate, but are in the region of $4000 compared with $2000), and enough must be purchased for the potential benefits to outweigh the additional complexity of management; for instance, by having enough to bundle together and to serve customers without requiring micromanagement. Furthermore, it is not simply the purchase price itself that is the issue; a high purchase price means that lessors will charge a higher rental price, meaning that they must be used intensively and not delivered on speculative routes where they may sit idle for a period of time before being required. This idle time is already a problem with regular containers; with a higher lease charge it would be unsustainable. These

Figure 2.4 Tank container. (From © TCCI [Own work] [CC BY-SA 3.0] via Wikimedia Commons.)

issues could be addressed by a pool of shippers purchasing their own containers, but that could only work on a regular loop back and forth between two destinations. This would involve additional costs and management, compared with regular containers, which are repositioned by shipping lines for any customer as required.

Container identification

Each shipping container has a unique identification. The previously mentioned ISO standard 6346 was established in 1995 to regulate the coding, identification and marking of the container. Referring to Figure 2.2 as an example, MSKU 0803081 45G1: The first three digits are the owner, in this case Maersk, and U is the product group code (U: freight containers, J: detachable freight container-related equipment, Z: trailers and chassis). The six digits are the registration number for the container, and the seventh is the check digit (an algorithm is used to calculate from the other six digits that should produce the correct check digit).

The four-digit combination of numbers and letters, for example 45G1, refers to the container type. ISO codes, both current (1995) and previous (1984), provide the length, height and width of the containers. The first figure in the four-digit container number records the length, the second figure records the height and width (a different figure for different height/width combinations, e.g. 5 is high cube while E or N is high cube with larger width), the third figure denotes the type (e.g. G for general container, R for reefer) and the fourth denotes the subset of that category (e.g. G0 is standard, G1 has vents). Therefore, for example, 45G1 is a 40 ft long container, 9 ft 6 in high, 8 ft wide (i.e. high cube and standard width), general-purpose type with vents (Tables 2.1 through 2.3).

Table 2.1 First digit marking for container length

Container length			
Mm	*Ft*	*in*	*ID*
2991	10		1
6068	20		2
9125	30		3
12192	40		4
7150			A
7315	24		B
7430	24	6	C
7450			D
7820			E
8100			F
12500	41		G
13106	43		H
13600			K
13716	45		L
14630	48		M
14935	49		N
16154			P

Table 2.2 Second digit marking for container height and width

Container height			*Container width*		
mm	*ft*	*In*	*2438 mm (8 ft)*	*2438–2500 mm*	*>2500 mm*
2438	8		0		
2591	8	6	2	C	L
2743	9		4	D	M
2895	9	6	5	E	N
>2895	>9	6	6	F	P
1295	4	3	8		
≤1219	≤4		9		

Table 2.3 Third and fourth digit for container type

General-purpose container without ventilation	
Opening(s) at one end or both ends	G0
Passive vents at upper part of cargo space	G1
Opening(s) at one or both ends plus 'full' opening(s) on one or both sides	G2
Opening(s) at one or both ends plus 'partial' opening(s) on one or both sides	G3
General-purpose container with ventilation	
Non-mechanical system, vents at lower and upper parts of cargo space	V0
Mechanical ventilation system, located internally	V2
Mechanical ventilation system, located externally	V4
Dry bulk container	
Non-pressurised, box type, closed	B0
Non-pressurised, box type, airtight	B1
Pressurised, horizontal discharge, test pressure 150 kPa	B3
Pressurised, horizontal discharge, test pressure 265 kPa	B4
Pressurised, tipping discharge, test pressure 150 kPa	B5
Pressurised, tipping discharge, test pressure 265 kPa	B6
Named cargo container	
Livestock carrier	S0
Automobile carrier	S1
Live fish carrier	S2
Thermal container	
Refrigerated, mechanically refrigerated	R0
Refrigerated and heated, mechanically refrigerated and heated	R1
Self-powered refrigerated/heated, mechanically refrigerated	R2
Mechanically refrigerated and heated	R3
Thermal container	
Refrigerated and/or heated, with removable equipment located externally; heat transfer coefficient K = 0.4 W/(m2*K)	H0
Refrigerated and/or heated with removable equipment located internally	H1
Refrigerated and/or heated with removable equipment located externally; heat transfer coefficient K = 0.7 W/(m2*K)	H2
Insulated; heat transfer coefficient K = 0.4 W/(m2*K)	H5
Insulated; heat transfer coefficient K = 0.7 W/(m2*K)	H6
Open-top container	
Opening(s) at one or both ends	U0
Opening(s) at one or both ends, plus removable top member(s) in end frame(s)	U1
Opening(s) at one or both ends, plus opening(s) on one or both sides	U2
Opening(s) at one or both ends, plus opening(s) on one or both sides plus removable top member(s) in end frame(s)	U3

(*Continued*)

Table 2.3 (Continued) Third and fourth digit for container type

Opening(s) at one or both ends, plus partial opening on one side and full opening on the other side	U4
Complete, fixed side and end walls (no doors)	U5
Platform (container)	
Platform (container)	P0
Fixed, two complete and fixed ends	P1
Fixed, fixed posts, either free-standing or with removable top member	P2
Folding (collapsible), folding complete end structure	P3
Folding (collapsible), Folding posts, either free-standing or with removable top member	P4
Open top, open ends (skeletal)	P5
Tank container	
For non-dangerous liquids, minimum pressure 45 kPa	T0
For non-dangerous liquids, minimum pressure 150 kPa	T1
For non-dangerous liquids, minimum pressure 265 kPa	T2
For dangerous liquids, minimum pressure 150 kPa	T3
For dangerous liquids, minimum pressure 265 kPa	T4
For dangerous liquids, minimum pressure 40 kPa	T5
For dangerous liquids, minimum pressure 60 kPa	T6
For gases, minimum pressure 910 kPa	T7
For gases, minimum pressure 220 kPa	T8
For gases, minimum pressure (to be decided)	T9
Air/surface container	
Air/surface container	A0

Empty container repositioning

In an ideal scenario, a loaded container would travel from origin to destination, where it would be stripped and then reloaded for export to a new destination. In practice, there is not always an export load waiting; therefore, once a container has been emptied, the empty box will be taken back to the nearest port or nominated depot. It may then wait there for a period of time until a local exporter requires it, or it may be sent back or 'repositioned' to the Far East, where most exporting is done. Around 30% of all recorded container handlings in world ports are empty containers. Western countries generally are net importers, meaning there are not enough export loads to fill all the containers that arrive with imported goods. Even if an export load is likely to be available, if the container must sit idle for more than 1–2 weeks then the loss of revenue becomes an issue, and the container owner would rather send the container to China where a load will definitely be found.

The problem arising from this system is that containers cost money to move, so the more empty or unproductive moves that take place, the higher the cost.

Initially, this cost is borne by the shipping line, but, particularly in difficult economic periods, this cost is often passed on to the shipper. It has been estimated that there exist about three containers for every container slot in the world fleet, to account for overland movements as well as taking up the slack in the system (Rodrigue, 2013). In 2008, at the peak of world container shipping, just before the recession, there were about 28 million TEU of containers in existence (UNCTAD, 2009). Most of these are controlled by shipping lines, either through ownership or by leasing them from container leasing companies, which provide flexibility for shipping lines that do not want to take the risk of purchasing too many containers. Shipping lines own approximately 62% of containers, and the remaining 38% is owned by leasing companies (Theofanis and Boile, 2009).

The problem with this system is that each container is owned (or at least controlled) by a separate shipping line. So, if an exporter is looking for empty equipment and locates some boxes owned by shipping line A at the nearby port, and if the exporter is a customer of shipping line B, then those boxes are not available to this exporter. The exporter will have to pay shipping line B to bring an empty container, while the empty boxes belonging to shipping line A may be unproductively repositioned elsewhere to serve shipping line A's customers. This results in additional movements and costs. There have been some attempts in the industry to solve this problem through the use of box pools (so-called grey boxes because containers are normally clearly branded for each shipping line), but the problem has not yet been resolved.

Swap bodies

Swap bodies can be moved between road and rail vehicles, but are not strong enough to be stacked or to be used on sea transport. They can be fully rigid or curtain sided for side loading. They often have the same external dimensions as general-purpose containers, but they possess four folding legs under the frame. As they do not have the same frame strength as containers, they cannot be lifted from above by spreader cranes but must be lifted from special fittings underneath. Special swap bodies may have more doors or sliding panels than ordinary hard boxes for trucks, railcars or sea containers, making unloading and loading faster and easier. Swap bodies are less strong than ISO containers, but they have some advantages as they are easier for truck drivers to connect to them as they do not have to be loaded and unloaded from ground level. Their less rigid construction also means that they have a lower tare weight (Figure 2.5).

Semi-trailers and chassis

A semi-trailer connects to a road tractor unit but it is also equipped with legs that can be lowered to support the trailer when it is uncoupled. The entire unit is called an *articulated lorry* or *articulated truck*, as opposed to a rigid vehicle

Figure 2.5 Swap body being loaded onto a rail wagon. (From Green Cargo.)

that is not able to pivot on the joint. Semi-trailers with two trailer units are called B-doubles or road trains. Like swap bodies, semi-trailers can be rigid or curtain sided or whatever formation is suitable for the cargo. Road vehicles can also be carried on rail wagons in their entirety (as in the Channel Tunnel). This is referred to as 'piggyback', and is less common than utilising a container (see Lowe, 2005; Woxenius and Bergqvist, 2011) (Figure 2.6).

The term *trailer* can also refer solely to the wheeled unit on which a container or swap body rests rather than an integrated loading unit. In the United States, the preferred term for a trailer is a *chassis*. In Europe, *grounded* intermodal terminals are the norm, whereby containers are transferred between trains and road trailers or stacked on the ground in between. In the United States, *wheeled* terminals are common, in which case containers are unloaded from trains onto waiting chassis, and the driver will arrive with only the tractor unit and hook up to a trailer or chassis. In Europe, the driver manages his or her own tractor unit and trailer and only the containers are interchanged.

Using a semi-trailer increases flexibility compared with a permanently coupled unit. Moreover, it can be combined with different types of tractors specifically

Figure 2.6 'Piggyback' semi-trailer being loaded on a rail wagon. (From Green Cargo.)

used for terminal operation and transportation. It also has a better ratio between its own and cargo weights.

The box trailer is the most common type of trailer, and it is also called a van trailer. A curtain sider is similar to a box trailer, except that the sides are movable curtains made of reinforced fabric covered with a waterproof coating. The purpose of a curtain sider is to allow the security and weather resistance of a box trailer with the loading ease of a flatbed. A regular curtain-sided semi-trailer is shown in Chapter 5. Figure 2.7 shows a curtain-sided intermodal container that can be transported by truck on a flat trailer or by rail on a flat wagon.

A reefer (or refrigerator truck) is a box trailer with a heating/cooling unit used to transport commodities requiring temperature control, such as cold chain products for supermarkets. They can also be compartmentalised for chilled, frozen or ambient, but current rail containers cannot, which limits their flexibility. A tanker is used for hauling liquids, such as gasoline, milk and so on. A dry bulk trailer resembles a big tanker, but it is used for sugar, flour and other dry-powder materials.

Double-deckers are trailers with a second floor to enable them to carry more palletised goods or retail cages. A standard lorry takes 45 retail cages, as does a 45 ft rail container, whereas a double-decker lorry can take 72 cages. As an example, double-deckers currently form about 20% of retailer Tesco's fleet in the United Kingdom (Monios, 2015). As a wholesaler, Costco uses pallets rather than cages, but they insist that all pallets are a maximum of 1.4 m high (with the occasional exception), so they can always stack them two-high in the trucks (Figure 2.8).

Figure 2.7 Tractor unit connected to curtain-sided intermodal container on a flat trailer. (From Rickard Bergqvist and Jason Monios.)

Figure 2.8 Double-deck trailer. (From Ray Forster on Flickr. 'Double-deck trailer' is copyright (c) 2013 Ray Forster and made available under a Creative Commons Attribution-NoDerivs 2.0 Generic License.)

Hinterland transport with containers and semi-trailers

The context of intermodal transport, and particularly for hinterland transport, of semi-trailers and containers is quite different. This section aims to describe those differences based on a number of characteristics, and it builds on the previous work of Woxenius and Bergqvist (2008, 2009, 2011). The context is described on the basis of the long-distance transport of containers and semi-trailers. The two segments of container- and semi-trailer–based transport are structured according to the main aspects of the logistics set-up, main markets, organisational structure and technology.

In terms of markets, semi-trailers serve mostly intra-regional flows, while the main transport market for maritime containers is the trans-ocean trade. The division is, however, not precise since the design of the latter transport system allows for co-production with intra-European container services, and the roll-on/roll-off (RoRo) ships transporting semi-trailers can also take containers on semi-trailer chassis.

The different transport services face different modal competition. Some RoRo services act as bridge substitutes with a clear sub-contractor role, while all-road or all-rail often constitute alternatives for semi-trailers in longer-range maritime services. Trans-ocean container services mainly compete with air transport although very differently in terms of costs and transport time, as analysed by Woxenius (2006). The development of trans-ocean container services has been driven by the growth in international trade to and from the Far East (cf. Woxenius and Bergqvist, 2011).

The business priority for the RoRo operators is mainly towards providing customer convenience, while the container segment aims at achieving economies of scale. The operational activity attracting most attention is, consequently, port operations for the semi-trailer segment and the maritime leg for the container

Table 2.4 Comparison between the container and semi-trailer shipping segments

Factor	*Container*	*Semi-trailer*
Geographic transport market	Trans-ocean/deep sea/ short sea	Intra-European/short sea
Modal competition	Air for deep sea leg Rail and road for feeder leg	Rail and road + fixed connections
Business priority	Utilising economies of scale	Providing customer convenience
Port geography	Few large hub ports + feeder ports	Many ports – partly bridge substitute
Hinterland depth	Deep	Shallow
Transport time/speed	Fast	Fast
Precision	Day	Hour
Order time	Week	Day/minute
Frequency	Weekly	Daily/hourly
Transport service coordinator	Shipping line, line agent or sea forwarder	Shipper, road haulier or general forwarder
Cargo dwell time in port	Days	Accompanied – minutes or none Unaccompanied – hours
Empty unit dwell time	Days/weeks	Hours/days
Port work content	Substantial	Limited
Rail technology	Very simple – flat wagon/ twist-locks	Complicated – pocket wagon/ king-pin box
Road technology	Awkward at end points	Simple and accessible
Road-rail transhipment technology	Fairly simple – automation possible	Dimensioning factor in weight and handling

Source: Woxenius, J. and Bergqvist, R., *Journal of Transport Geography*. 19(4), 680–688, 2011.

segment. The results are quick RoRo transhipment and frequent departures versus lift-on/lift-off (LoLo) transhipments and hub-and-spoke systems and well-planned capacity. In other words, the focus of RoRo can be characterised as being on service, while that of LoLo is on low transport costs (Woxenius and Bergqvist, 2011).

The general operating characteristics are that the container segment operates through a hub-and-spoke system with a relatively small number of hub ports combined with feeder services to regional ports, while the lower dependence on economies of scale in the RoRo segment has led to maintained service in a straight line with less focus on large gateway ports.

The modal competition has led to a sharper geographic concentration in the container segment, implying larger hinterland depth for containers; that is, they generally travel further inland from each port than the semi-trailers do.

The RoRo segment presents a wider range of possible time-schedule changes in order to align the number of turnarounds between individual routes. Shippers would expect time precision in hours or even minutes for the semi-trailer segment, while the container segment is less strict on time precision (maybe days instead of hours). The frequency is often much higher in the RoRo segment, where daily sailings are customary for unaccompanied semi-trailer traffic and services that act as bridge substitutes can often have very frequent services.

The dwell time in ports for semi-trailers is often very short compared with the container segment, where ports are often used to absorb the slack in the transport planning chain or to manage capacity gaps between the container ship and the vehicles used in land traffic modes. The fact that semi-trailers are of higher value also contributes to the low dwell time of empty units.

Container transport is typically organised by the shipping lines, their agents or specialised sea forwarders. For them, it is common practise to think in transport chains that split between modes, whereas the road hauliers and the transport of semi-trailers normally plan for the same vehicle throughout the transport chain. The planning and operational barriers for using rail are accordingly higher for semi-trailers.

The physical characteristics of the container and the semi-trailer evidently affect the technology that surrounds them. At the same time, much of the transport technology is multipurpose in the sense that it can manage both types of load units.

In general, the employed rail technology, as well as the transhipment technology, is more complicated and costly for semi-trailers than for containers. The height and weight of the semi-trailer require the use of four-axle pocket wagons. In some instances, the limited rail loading profile hinders the rail transport of semi-trailers, as in parts of the rail network in the United Kingdom for example. The different characteristics of the container and semi-trailer shipping segments are summarised in Table 2.4. It is obvious that semi-trailers meet very different preconditions for (hinterland) transport compared with containers.

HANDLING EQUIPMENT

Forklifts

There are many types of forklifts, which vary according to their functionalities and the specifications of the manufacturers. They are used mostly in the warehouse or terminal area. In warehouses, they are used for handling all kinds of palletised goods, sacks, drums and so on. In container terminals, forklifts tend to specialise in moving empty containers (Figure 2.9).

An important aspect of forklift operation is that many have rear-wheel steering. This increases manoeuvrability in tight cornering situations. However, one criticism is its instability. The forklift and load must be considered a unit with a continually varying centre of gravity with every movement of the load. A forklift must never negotiate a turn at speed with a raised load, where centrifugal and gravitational forces may cause the truck to tip over.

Figure 2.9 Forklift handling an empty container. (From Michael Coghlan on Flickr. 'Handling a Container' is copyright (c) 2011 Michael Coghlan and made available under a Creative Commons Attribution-NoDerivs 2.0 Generic License.)

Reach stackers

A reach stacker is one of the most popular pieces of handling equipment for container handling operation in a small- or medium-sized terminal and port. This equipment is very flexible. It is able to transport a container over short distances very quickly and stack them in various rows, depending on its access. Reach stackers are a much more cost-effective option for small- to medium-size terminals, compared with the major investment required for a gantry crane (Figure 2.10).

Rubber-tyred gantry cranes (RTG)

Rubber-tyred gantry cranes (RTG) are equipment for the yard handling of standard containers. They are a very common part of handling systems in large ports, container terminals and container storage yards. RTGs are an efficient solution when straddling multiple lanes of containers. Their advantage over rail-mounted gantry cranes (RMG) is that they can be moved to different locations more easily (Figure 2.11).

Figure 2.10 Reach stacker loading a container onto a train. (From Rickard Bergqvist and Jason Monios.)

Figure 2.11 Rubber-tyred gantry crane. (From Rickard Bergqvist and Jason Monios.)

Figure 2.12 Rail-mounted gantry crane. (From Rickard Bergqvist and Jason Monios.)

Rail-mounted gantry cranes (RMG)

Rail-mounted gantry cranes (RMG) travel on rails to lift and stack containers in the yard area using a spreader (or twin-lift spreader, if needed). They come in a variety of models with different spans and overhangs. Compared with the RTG, the RMGs have the advantages of being powered by electricity, greater lifting capacity and higher travelling speed. RMG have proved particularly effective for rail/road transhipments of large quantities, since the cranes required for this purpose must be able to move quickly in both a longitudinal and a transverse direction. They are also commonly used in large ports to handle containers in stacks, while ship-to-shore (STS) gantry cranes move containers between the vessel and the quay (Figure 2.12).

STS cranes

STS cranes are large gantry cranes that are used in ports for loading and unloading vessels at the quayside. They have a higher and longer reach and faster handling time. A large container port will use a number of cranes simultaneously to handle a large container vessel. See Chapter 7 for more discussion (Figure 2.13).

Straddle carriers

Straddle carriers drive astride containers or swap bodies to lift, carry or stack as required. They are used in ports to move containers between the

Figure 2.13 STS cranes unloading a container vessel. (From Port of Gothenburg.)

Figure 2.14 Straddle carrier. (From Port of Gothenburg.)

loading/unloading areas and the container stack. See Chapter 7 for more discussion (Figure 2.14).

Automated guided vehicles

An automated guided vehicle (AGV) follows markers or wires under the ground or uses vision or lasers. They are most often used in industrial applications to move materials around a manufacturing facility or a warehouse. AGVs are used in some ports to produce a fully automated container terminal (Figure 2.15). Due to the high investment and high maintenance required, the systems are

Figure 2.15 AGVs in operation at a port container terminal. (From © Henrik Jessen [CC BY 3.0] via Wikimedia Commons.)

still seen as not as competitive as the conventional systems such as straddle carriers, for most of the container terminal operators. See Chapter 7 for more discussion.

Vehicle/trailer-mounted loading equipment

Mobile loading equipment is necessary for small terminals that do not have the funds to make investments in expensive equipment, or for customers that want to rent equipment for a short-term need and an intermodal terminal is not available. For example, some warehouses may not have raised loading bays or raised storage and need the driver to place the container on the ground, or small intermodal terminals with very little traffic can load trains directly in this manner without needing a reach stacker.

TRAINS

Locomotives

There are many characteristics of locomotives. The purpose of this section is to briefly introduce the main types and most commonly used locomotives. The locomotive itself has no payload capacity but is purpose-built for the traction of rail wagons. For the transport of freight, only isolated locomotive power is used, as opposed to passenger transport where the engine might be integrated into the passenger wagon in what is called a ***multiple unit***. The advantage is that the locomotive is easily exchangeable, meaning that the locomotive can easily move from one freight assignment to another by just switching wagons. Locomotives can be divided into three main types: electric, non-electric and hybrids. The classification of non-electric includes fuels such as steam, gasoline and diesel. The hybrid category contains locomotives where a diesel engine is combined with an electric engine.

Besides locomotive design for freight transport, there are also locomotives designed for the shunting and marshalling of wagons at terminals. These locomotives do not need to be designed for high speed but can operate at much lower speed. They can therefore be designed more compactly than regular freight locomotives. Shunting locomotives are usually also remote controlled to facilitate operations and the driver's opportunity to oversee the operations without the need to sit inside the locomotive driver's compartment. Both of them need to be designed to provide enough traction to move several thousand tonnes of freight. A regular freight locomotive used in Europe would normally have the capacity of up to 4000 kW (equivalent to about 5400 horsepower). Locomotives that move goods such as raw materials (e.g. iron ore) on rail lines where there is high axle load capacity can be equipped with engines of more than 10,000 kW in power (Figure 2.16).

The first locomotives were steam powered, and they gave way to diesel midway through the twentieth century. In the last few decades, electric locomotives have become increasingly common. Electric locomotives are the most energy efficient and provide the most cost-efficient operation. Many different types have been developed over the years, but the principle is more or less the same. The electric lines on rail networks operate with very high voltage in order to enable many locomotives to run at the same time. The electricity goes from the line via the locomotive connector (see Figure 2.17) to a transformer on the locomotives, where the voltage is lowered. The transformer has several outlets, which are managed by the driving controls. When the driver adjusts the speed, more or fewer outlets are connected, and thus the voltage to the engine changes as does the speed of the locomotive. Through a transmission, the power is transferred to the blind-driving wheel, and coupling rods connect it with the rest of the driving

Figure 2.16 Freightliner electric locomotive. (From Rickard Bergqvist and Jason Monios.)

Figure 2.17 Electric locomotive showing location of power-connecting mechanism. (From Green Cargo.)

wheels. It is important to note that the emission profile of electric locomotives depends on the method used to generate the electricity. If the electricity was derived from a coal-fired power station, then the rail movement may not be any greener than using diesel (see Chapter 10).

The drawback of electric locomotives is the need for infrastructure to provide the power along the entire network. In countries with very large networks involving long distances (e.g. North America), the cost is too prohibitive; therefore, diesel locomotives remain the norm. It is also common to use diesel locomotives in terminals as overhead infrastructure because electricity connection would be in the way of handling equipment.

One of the most common types of diesel-powered locomotives is the diesel-electric (see Figure 2.18). The diesel engine produces electricity through a generator. The engine output is regulated by the driving controls and transferred to the electric engines, normally one at each axle, mounted inside the bogie.

Figure 2.18 Diesel-electric locomotive. (From Green Cargo.)

Besides engine performance, wheel arrangement is also very important for the overall effectiveness of locomotives. The more wheels that can transfer the power, the more traction that can be transferred to the track, hence the more robust the locomotive becomes to conditions such as a slippery track caused by leaves on the tracks, ice and so on. For slippery tracks, many locomotives are also equipped with sanding equipment that can put a small amount of sand on tracks to improve friction.

Different systems exist for describing and illustrating how the axles are arranged. The most common is the International Union of Railways (UIC, 1983) classification. It contains several dimensions, but the most important is the number of consecutive driving axles (starting at A for a single axle, B for two axles, etc.). Besides the letter, the number following it describes the consecutive non-driving axles, starting from 1 for a single axle. The prime sign (´) illustrates if the axles are mounted on a bogie. The lower-case letter 'o' indicates that each axle has its own engine (individually powered). Here are some examples of common configurations (see Figure 2.19):

- B´B´. Two bogies. Each bogie has two driving axles.
- Bo´Bo´. Two bogies where each bogie has two individually powered axles.

B´B´ and Bo´Bo´ cover the majority of configurations of modern locomotives. Age, engine, type of power, axle arrangements, configurations and so on, all exert a significant influence on the cost of locomotives. Investments in locomotives are very capital intensive and there are many types of locomotives on the market, old and new. Since locomotives are so expensive, it is not unusual for old locomotives (up to 30–40 years old) to still be in use for regular freight transportation. It is difficult to give an exact estimate of the cost of locomotives due to the large

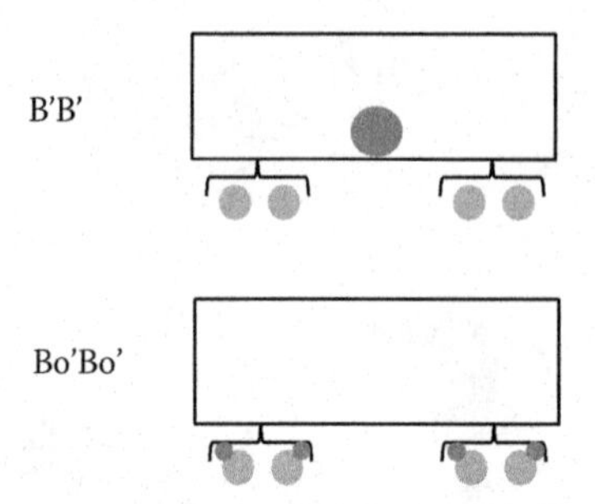

Figure 2.19 Illustration of axle arrangement B′B′ and Bo′Bo′ (UIC classification). (From Rickard Bergqvist and Jason Monios.)

difference in types and ages; however, a modern locomotive can be expected to cost around €1–3 million depending on the power and technology used.

Wagons*

In recent decades, flat wagons for hauling containers have become the dominant type of wagon, replacing the boxcar, where the cargo had to be loaded and unloaded into the wagon manually. Yet, many specialised kinds of wagons continue to be used. Other specialist equipment permits a semi-trailer to be driven onto the train, allowing for easy transhipment between road and rail.

There are numerous types of goods wagons, categorised according to the UIC classification system, which was first implemented in 1965 and revised over time as new wagon types entered service (Table 2.5). While many types of wagons are used, they typically have standardised twist-locks and other fittings such as hoses for air brakes.

Open wagons (referred to as gondolas in the United States) specialise in bulk goods, such as coal or iron ore, which are the traditional traffic type for rail freight transport. Open wagons of standard design (UIC Class E) often have side doors but without self-discharging equipment (Figure 2.20). Open wagons may also be of special design (UIC Class F). The common type is a hopper wagon, which is self-discharging and thus used where automated handling equipment can fill and empty the wagons. A common example is a coal hopper wagon used at specialist coal ports (see Figure 2.21).

Covered wagons or vans (boxcars in the United States) have a fixed roof and are less common now that intermodal transport of consumer goods uses mostly ISO containers. These wagon types thus specialise in part-load goods or parcels. They are classified as ordinary class (UIC Class G – see Figure 2.22) or special class (UIC Class H). Refrigerated wagons (Class I wagons) are similar to the refrigerated ISO containers mentioned earlier in this chapter. They are built with

* Note: the term *wagon* is used in Europe, whereas *car* is preferred in the United States.

Table 2.5 UIC wagon codes

Class	*Wagon type*
E	Ordinary open high-sided wagon
F	Special open high-sided wagon
G	Ordinary covered wagon
H	Special covered wagon
I	Refrigerated van
K	Ordinary flat wagon with separate axles
L	Special flat wagon with separate axles
O	Open multipurpose wagon (composite open high-sided flat wagon)
R	Ordinary flat wagon with bogies
S	Special flat wagon with bogies
T	Goods wagon with opening roof
U	Special wagons
Z	Tank wagon

Figure 2.20 Open wagon with tarpaulin cover. (From Wascosa.)

insulation for temperature control and require a power source either built in or plugged into the train's power.

Wagons with a sliding roof (UIC Class T) either have a flat wagon floor or equipment for self-discharging. The special wagons of UIC Class U include powder wagons and low-loading wagons (Figure 2.23). Tank wagons (UIC Class Z) are suitable for a wide variety of fluids and gases (see Figure 2.24). Goods wagons for special purposes include works wagons, used by railway administrations exclusively for their own internal works purposes; ferry wagons with smaller loading gauges for traffic travelling to the United Kingdom, which were designated with a lower-case letter f; and the rarely mixed open, flat wagons of UIC Class O, which are equipped with folding sides or stakes and can be used either as flats or as open goods wagons.

Flat wagons (flatcars in the United States) have no walls or low walls that are no higher than 60 cm. The kind with uprights can be used for goods, such

Figure 2.21 Open wagon. (From © Phil Sangwell [CC BY 2.0] via Wikimedia Commons.)

Figure 2.22 Covered wagon being unloaded by a forklift. (From Port of Gothenburg.)

as timber logs and steel pipes (Figure 2.25). Flat wagons without any walls or uprights are the common wagons for the intermodal transport of ISO containers, often referred to in the United States as container on flatcar (COFC), to distinguish from trailer on flatcar (TOFC) (see Figure 2.26).

Figure 2.23 Powder wagon. (From Wascosa.)

Figure 2.24 Tank wagons. (From Port of Gothenburg.)

Figure 2.25 Flat wagon with uprights for hauling general cargo such as steel pipes. (From Port of Gothenburg.)

Figure 2.26 A flat wagon carrying a 40 ft ISO container. (From Rickard Bergqvist and Jason Monios.)

Flat wagons are very widely used, since it is convenient to handle and transfer the whole container from the trailer directly onto the train. Different rail networks will have their own common wagon sizes; for example, in some countries, 80 ft wagons are popular for combining 20 and 40 ft containers. In the United Kingdom, 60 ft wagons are common to handle multiples of 20 and 40 ft containers moving through ports. Fifty-four-foot wagons are often used in the United States to haul 53 ft containers, and they are also used in the United Kingdom for carrying 45 ft wagons – an inefficient use of limited train length (Figures 2.27 and 2.28).

Figure 2.27 A 45 ft container on a 54 ft wagon. (From Rickard Bergqvist and Jason Monios.)

Figure 2.28 Two flat wagons coupled together and showing wagon markings with technical information on weight limits. (From Rickard Bergqvist and Jason Monios.)

Figure 2.29 Well car with double-stacked 53 ft US containers. (From Rickard Bergqvist and Jason Monios.)

In some countries, double stacking containers is possible according to the clearance of tunnel and bridge heights (Figure 2.29). This is common in the United States and is becoming more widely used in countries such as China.

CONTAINER VESSELS

Deep-sea container vessels

Table 2.6 shows the entire world fleet of seagoing container vessels. Vessels range from only a few hundred TEU to the current largest size of around 20,000 TEU. TEU capacity is somewhat nominal as it depends on whether the containers are full or empty. Over time, maritime networks have developed from direct calls between ports to a hub-and-spoke system, whereby increasingly large vessels transport thousands

Table 2.6 World cellular fleet January 2016

	In service Jan 2016		*On order 2016*		*On order 2017*		*On order 2018+*			
TEU range	*No.*	*TEU*	*No.*	*TEU*	*No.*	*TEU*	*No.*	*TEU*	*Total vessels on order*	*Total TEU on order*
0–499	319	82,883	10	2,986	0	0	0	0	10	2,986
500–999	688	523,162	6	4,513	0	0	0	0	6	4,513
1,000–2,999	1,852	3,337,838	90	175,010	75	145,360	43	94,436	208	414,806
3,000–4,999	915	3,779,536	5	20,800	11	38,700	12	43,836	28	103,336
5,000–7,499	628	3,785,623	2	13,400	5	26,500	3	15,900	10	55,800
7,500–9,999	430	3,736,165	34	311,877	4	36,400	0	0	38	348,277
10,000–12,999	101	1,087,346	19	198,020	25	273,000	16	188,800	60	659,820
13,000–15,999	170	2,305,604	25	353,000	23	323,500	28	398,750	76	1,075,250
16,000+	48	870,575	13	246,443	26	525,480	29	582,500	68	1,354,423
Total	5,151	19,508,732	204	1,326,049	169	1,368,940	131	1,324,222	504	4,019,211

Source: Containerisation International.

of containers between hubs, and then hundreds or thousands of containers are transhipped onto smaller feeder vessels to be transported to smaller ports, and vice versa. Table 2.6 shows that there is still a lot of capacity in the mid-range of container vessel sizes, but the average size is increasing as more ultra-large container vessels are ordered.

Container vessels are fitted with cell guides where containers of fixed lengths (20 or 40 ft) can be attached via twist-locks and corner castings, with more containers stacked on top. They are then further secured with lashing rods. These all need to be secured and released manually by dock workers. Some specialist trades have different size fittings, for example short sea intra-European vessels are designed to accommodate 45 ft long containers.

It has been estimated that a 19,000 TEU vessel dropping 8,800 TEU in a single call will necessitate 14,000 container moves, six 800 TEU feeders, 53 trains (carrying 90 containers each), three 96 TEU barges and 2,640 trucks (Grey, 2015). Due to these difficulties, the newest generation of container ships inherently represent a greater risk for the owner, be they a lessor or operator, as the cost of around $150+ million for one vessel is a significant outlay that must be recouped. Figure 2.30 shows a large container vessel. See Chapter 7 for full discussion of ports and container vessels.

Feeder and short sea container vessels

Feeder vessels are the same type as deep-sea container vessels but with smaller capacities to suit feeder or spoke movements from hub ports to smaller ports that do not have sufficient demand to have their own direct calls. Many of these can also traverse ocean routes depending on weather conditions, so what is classed a feeder vessel is somewhat relative to trade requirements. There is some evidence that very small vessels may be phased out (see differential in numbers of orders

Figure 2.30 Large container vessel. (From Port of Gothenburg.)

for small vessels in Table 2.6) due to more attractive economies of scale achievable from larger vessels.

Short sea, as opposed to *feeder*, designates a direct flow of traffic on a point-to-point routing, that is, direct trade rather than cargo transhipped at an intermediate port. This terminology is commonly used in Europe to differentiate between intra-European traffic and global traffic being transhipped at European ports and then feedered to small ports. The main reason for the distinction is that intra-European flows use 45 ft containers (in order to achieve better container fill comparable to what is achieved on a 45 ft truck trailer). As these are set up with cell guides for 45 ft containers, they cannot accommodate 20 ft/40 ft containers transported on other routes; therefore, they specialise in their own market. But there is no difference in the vessels themselves (Figure 2.31).

Inland navigation vessels

Barges are flat-bottomed, open-deck vessels, normally towed or pushed by a tug, although some barges have their own engines (see Figure 2.32). As canals are sheltered from the sea, barges do not require the same seaworthiness as short sea or sea-inland vessels, although their use on rivers will depend on the local geographical conditions and the level of exposure to coastal weather conditions. They range in capacity from around 32 to 48 TEU to larger vessels up to over 400 TEU. See Chapter 4 for more discussion of inland navigation (Figure 2.33).

Figure 2.31 Feeder container vessel. (From Port of Gothenburg.)

Figure 2.32 Containers being loaded into a container barge. (From Rickard Bergqvist and Jason Monios.)

Figure 2.33 Container barge on the Yangtze River. (From Rickard Bergqvist and Jason Monios.)

Roll-on/roll-off

A roll-on/roll-off vessel, or RoRo, is a type of ship designed to carry wheeled cargo, such as automobiles, trucks, semi-trailer trucks, trailers and railroad cars that are driven on and off the ship via a built-in ramp for efficient loading and discharging operations. RoRo vessels allow semi-trailer trucks to drive directly into the ship, where they can be transported in their entirety or where the semi-trailer may be left unaccompanied. Even containers can be left at the port and placed on dollies or mobile loading platforms (MAFIs) and wheeled into the vessel in that way, and then picked up by tractor units at the unloading point. Accompanied, full, articulated lorries are popular for short crossings, whereas longer crossings are more suitable for unaccompanied transport as there is no need to pay for a driver to be with the vehicle for many 'stationary' hours. This type of operation is often seen in the North Sea and the Baltic Sea region (Figure 2.34).

Figure 2.34 RoRo vessel with ramp for vehicle access. (From Port of Gothenburg.)

CONCLUSION

This chapter has provided an overview of the main vehicle and equipment types used in intermodal transport. However, due to the vast variety of equipment in use in the industry, the specific types described here are only representative of some of the more commonly seen varieties. Students are encouraged to go online to view photographs and videos through which to explore the wide variety of equipment in addition to that listed in this chapter, and in particular to view how it is used in practice, such as in loading and unloading procedures.

REFERENCES

Bouley, C. (2012). Manifesto for the 45' palletwide container: A green container for Europe. Available at http://issuu.com/cjbouley/docs/manifesto_for_the_45_pallet_wide_container. Accessed 16 March 2012.

Grey, M. (2015). Age of the giants. *Lloyd's List*. Available at http://www.lloydslist.com/ll/sector/containers/article456093.ece. Accessed 16 March 2015.

Konings, R. (2005a). Foldable containers to reduce the costs of empty transport? A cost-benefit analysis from a chain and multi-actor perspective. *Maritime Economics & Logistics*. 7: 223–249.

Konings, R. (2005b). How to boost market introduction of foldable containers? The unexpected role of container lease industry. *European Transport\Trasporti Europei*. 25–26:81–88.

Levinson, M. (2006). *The Box: How the Shipping Container Made the World Smaller and the World Economy Bigger*. Princeton University Press, Princeton, NJ.

Lowe, D. (2005). *Intermodal Freight Transport* Elsevier Butterworth-Heinemann, Oxford.

Monios, J. (2015). Integrating intermodal transport with logistics: A case study of the UK retail sector. *Transportation Planning and Technology*. 38 (3): 1–28.

Monios, J., Wilmsmeier, G. (2014). The impact of container type diversification on regional British port development strategies. *Transport Reviews*. 34 (5): 583–606.

Murdoch, E., Tozer, D. (2012). *A Master's Guide to Container Securing*. Available at http://www.standard-club.com/media/24168/AMastersGuidetoContainerSecuring2ndEdition-3.pdf. Accessed 6 April 2017.

Rodrigue, J.-P. (2013). The repositioning of empty containers. Available at http://people.hofstra.edu/geotrans/eng/ch5en/appl5en/ch5a3en.html. Accessed 29 May 2013.

Shintani, K., Konings, R., Imai, A. (2012). The effect of foldable containers on the costs of container fleet management in liner shipping networks. *Maritime Economics & Logistics*. 14: 455–479.

Theofanis, S., Boile, M. (2009). Empty marine container logistics: Facts, issues and management strategies. *Geojournal*. 74 (1): 51–65.

UIC. (1983). UIC Leaflet 650. Standard designation of axle arrangement on locomotives and multiple-unit sets. 5th edition of 1-1-1983.

UNCTAD. (2009). *Review of Maritime Transport*. Geneva: UNCTAD.

Woxenius, J. (2006). Temporal elements in the spatial extension of production networks. *Growth and Change*. 37 (4): 526–549.

Woxenius, J., Bergqvist, R. (2008). Hinterland transport by rail: A success for maritime containers but still a challenge for semi-trailers. Conference proceedings, Logistics Research Network Annual Conference, 10–12 September 2008, University of Liverpool, UK.

Woxenius, J., Bergqvist R. (2009). Hinterland transport by rail: Comparing the Scandinavian conditions for maritime containers and semi-trailers. *International Association of Maritime Economists (IAME) Conference*, 24–26 June 2009, Copenhagen, Denmark.

Woxenius, J., Bergqvist R. (2011). Hinterland transport by rail: Comparing the Scandinavian conditions for maritime containers and semi-trailers. *Journal of Transport Geography*. 19 (4): 680–688.

Chapter 3

Rail operations

Allan Woodburn

INTRODUCTION

This chapter presents the key aspects of intermodal freight transport as they apply to rail activity. The focus is primarily on rail operations themselves, though with some consideration of the interface between train and terminal. Intermodal techniques offer a means by which rail can broaden its coverage of the contemporary freight market by replicating some of the characteristics of the bulk traffic for which it is typically well-suited. This is particularly important in deindustrialising economies (notably the United Kingdom, but also other European countries) where there has been a steep decline in bulk flows associated with the manufacturing sector and a corresponding increase in the share of the freight market accounted for by consumer goods. The more dispersed nature of consumer goods flows compared with bulk flows makes it less likely that suppliers and customers will have direct rail connections to their premises, so unitising the goods increases the likelihood of rail becoming a viable modal option.

The level of penetration of intermodal rail freight in the overall rail freight market varies considerably from country to country and different countries have different mode shares for rail in their total freight market. Limited data relating specifically to intermodal rail freight activity in most countries poses a challenge when trying to understand market trends, but it is evident that some countries have witnessed considerable growth in intermodal rail freight activity in recent years while the markets of others have remained relatively stable. The United Kingdom is one country whose intermodal rail freight market has grown rapidly over the last 20 years; where appropriate, this chapter incorporates original research findings related to British intermodal rail freight operations.

The emphasis in this chapter is on unaccompanied intermodal transport, where the unit load travels by rail on its own, since this forms the overwhelming majority of intermodal rail freight activity. Some consideration is also given to the specific characteristics of accompanied intermodal transport, conveying entire road vehicles accompanied by their drivers. Within the European context, it is important to note that much of the published information, including statistics, relates to the rail element of combined transport, which is not entirely synonymous with intermodal transport. Critically, in the official definition of combined transport, the rail leg of an intermodal journey must form the majority of the distance and must exceed 100 km (Council of the European Union, 1992). Some important flows, such as the lorry shuttles through the Channel Tunnel, do not satisfy these criteria and are omitted from most intermodal activity statistics. In other cases, statistics only include the activity of member organisations and do not reflect the entirety of the intermodal rail freight market. As far as possible, this chapter adopts an all-encompassing approach to understanding the topic.

By way of context, prior to considering specific aspects of rail operations, the following sections briefly identify the key intermodal rail freight markets and the different network design concepts. The next section provides the rail network infrastructure and train operations perspectives, followed by an account of the rail equipment considerations. The basics of intermodal rail freight economics are then presented, before addressing the ways in which accompanied intermodal transport differs from the more common unaccompanied mode of operation. The penultimate section briefly considers the future direction of intermodal rail freight and a short conclusion to the chapter is provided in the final section.

INTERMODAL RAIL FREIGHT MARKETS

Fundamentally, intermodal rail freight activity falls into one of two categories: port-hinterland (or maritime) flows and continental/domestic (or non-maritime) traffic. This distinction is reflected in much of the literature, from both the industry and academic perspectives. In some cases, intermodal services carry unit loads for both markets, so the distinction between the two markets is not absolute in practice. A detailed analysis of the European intermodal rail freight market, including some comparative information for North America, can be found in a report prepared for the European Commission (2015). Figure 3.1 shows the evolution of the rail component of European intermodal transport activity from 1990 to 2014. Total activity increased by 145% during this time period, though the trend has been volatile since 2007, when the global financial crisis began.

Figure 3.2 specifically considers the evolution of intermodal rail freight in Britain since, as mentioned earlier, there has been considerable growth in recent years. Between 1998/1999 and 2015/2016, intermodal volumes increased by 82% at a time when the total rail freight volume grew by just 2.4% (ORR, 2016). As a consequence, the percentage of British rail freight activity accounted for by intermodal increased from 20% to 36% during this period.

Table 3.1 sets out the extent of the different road-rail combined transport market segments in the European Union (EU), with international flows accounting for more than half of total activity. The international market segment includes direct rail flows across EU borders to/from other countries, but is dominated by the European hinterland rail flows of the International Organization for Standardization ISO containers passing through EU ports and feeding the global shipping industry. The other two market segments essentially comprise domestic

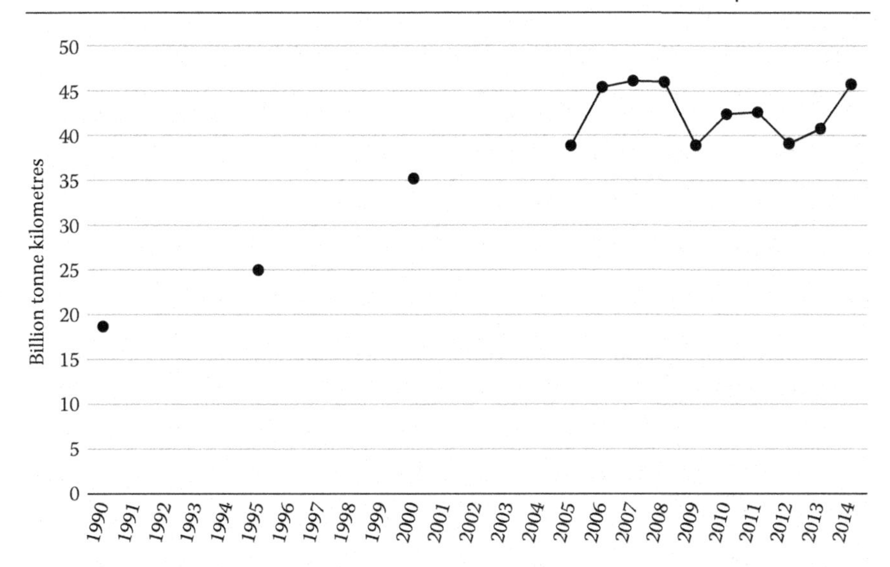

Figure 3.1 Evolution of the rail component of European intermodal transport, 1990–2014. (Based on UIRR, UIRR Report: European Road-Rail Combined Transport 2014–15, International Union for Road-Rail Combined Transport [UIRR], Brussels, 2015.)

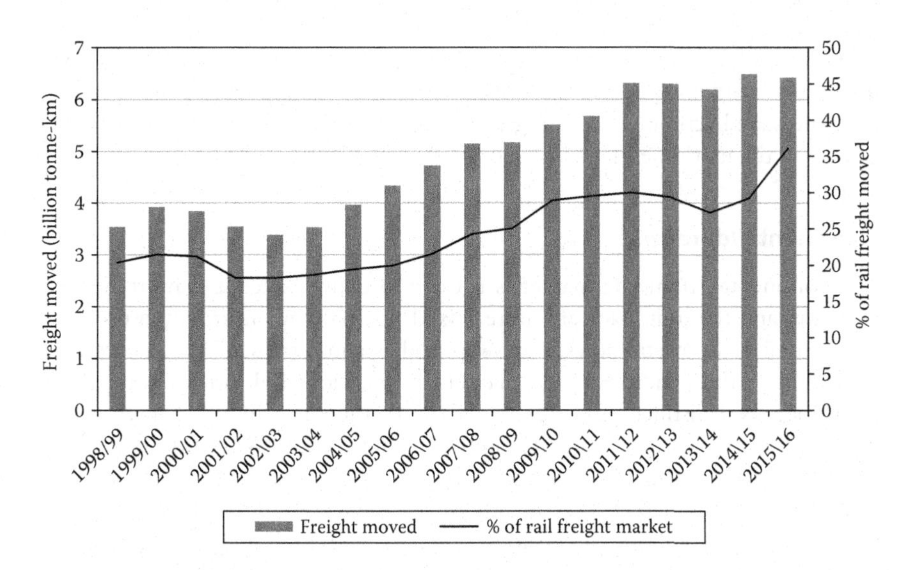

Figure 3.2 Domestic intermodal rail freight moved and share of rail freight market, 1998/1999–2015/2016. (Based on ORR, Data portal: Freight rail usage, https://dataportal.orr.gov.uk/browsereports/13, 2016.)

Table 3.1 Total road-rail combined transport activity in the EU, by market segment (2011)

Market Segment	*TEU Volume (m)*	*Estimated Tonnes Moved (bn rail t-km)*
Intra-member state (MS)	3.218	17
Intra-EU	4.856	51
International (i.e. to/from EU)	9.134	47
Total	17.208	115

Source: Based on European Commission, Analysis of the EU Combined Transport: Final Report, Prepared by KombiConsult GmbH, Intermodality Ltd, PLANCO Consulting GmbH and Gruppo CLAS S.p.A., Brussels: European Commission, 2015.

(i.e. intra-MS) and continental (i.e. international but intra-EU) flows. Both markets are discussed in the following sections.

Port-hinterland (maritime)

The characteristics of the port-hinterland market are heavily influenced by the deep-sea container shipping sector, where rail plays a supporting role for the land-borne movement of ISO containers between ports and inland terminals. In many respects, this is the more straightforward of the two categories since the concentration of volume and the standardised nature of ISO containers lead to considerable conformity of rail operations. This is particularly the case for deep-sea traffic, where containers of 40' length are dominant along with a sizeable proportion of 20' containers. Where rail is handling short-sea intermodal flows, for example linking mainland Europe, the United Kingdom and Ireland, 45' containers are becoming increasingly popular. However, the vast majority of rail activity in the maritime market is associated with deep-sea volume.

Continental/domestic

The continental/domestic market is trickier to define, since it is more diverse in nature and the unit loads are more mixed. ISO containers are often used for continental or domestic flows, but swap bodies and lorry semi-trailers are also important in the unaccompanied market. As mentioned already, accompanied intermodal is a distinct sub-category of the continental/domestic market and its characteristics are discussed separately in a later section. While unit load characteristics in the maritime market are dictated by deep-sea shipping, in the continental/domestic market, they mostly reflect the variety of road-based activity rather than the more rigid dimensions associated with maritime container traffic. The greater variability in unit load type and length has particular implications for wagon design and on-train utilisation, issues that are discussed later in this chapter. With the ongoing attempts to improve transport efficiency, particularly in relation to cubic capacity since this is increasingly the constraining factor rather than weight, larger unit loads have become more commonplace, generally where they are also permitted to travel by road. For example, the Malcolm Group (a British logistics service provider which makes considerable use of intermodal rail) has introduced 50' containers to its domestic intermodal routes (Malcolm Group, 2014). In North America, 48' and 53' containers are commonly used for rail-road intermodal flows.

Niche intermodal rail freight

It is worth pointing out that intermodal techniques are sometimes used for niche rail freight traffic, which in many respects resembles bulk traffic. In the United Kingdom, for example, flows of domestic waste and gypsum move in dedicated ISO containers in trainload volumes. Such flows are not recorded in British intermodal statistics, but they use the same principles and can use the same handling equipment and wagons as standard intermodal traffic. Elsewhere in Europe, companies such as Innofreight (http://www.innofreight.com/en/) have developed bespoke containers and associated equipment for a wide range of bulk goods, such as coal, scrap metal, construction materials and agricultural produce. Again, these flows almost always operate independently of the more standard intermodal rail flows on which this chapter focuses.

RAIL NETWORK DESIGN FOR INTERMODAL OPERATIONS

First, general network design concepts are established, followed by a case study of the intermodal markets and network strategies of the British rail freight operators, demonstrating a practical example of network implementation.

General network design concepts

A key consideration for intermodal rail freight is whether to integrate intermodal traffic with other freight flows or to operate dedicated intermodal services. Reliable statistics are not available but, on the whole, dedicated provision is dominant. It is not uncommon to see combined intermodal and traditional rail freight jointly carried on freight trains in countries where the shared-user wagonload provision still forms a sizeable proportion of rail freight activity, but even then the vast majority of long-distance intermodal traffic is conveyed on dedicated trains. The left picture in Figure 3.3 illustrates a dedicated intermodal service in Germany, in this case conveying maritime containers. The right picture, from Austria, shows a shared-user train conveying both traditional rail wagons and intermodal wagons loaded with containers. As will be seen later in this chapter, it is generally sensible to provide dedicated services where this is a viable option.

Figure 3.3 Examples of dedicated intermodal service provision (left picture) and shared service provision including intermodal (right picture). (From Allan Woodburn.)

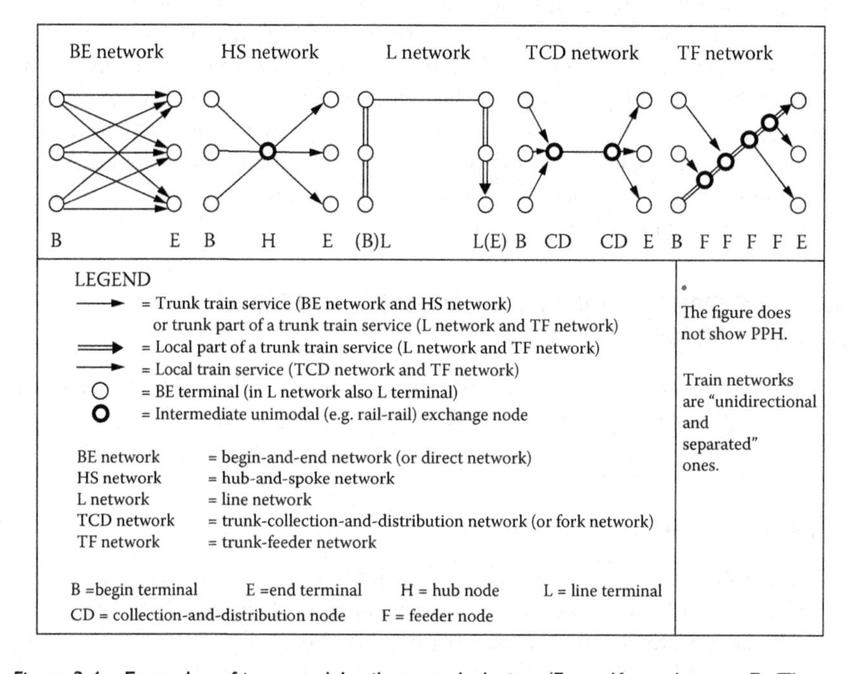

Figure 3.4 Examples of intermodal rail network design. (From Kreutzberger, E., The innovation of intermodal rail freight bundling networks in Europe: Concepts, developments, performances, PhD thesis, Delft University of Technology, Delft, 2008.)

There is a considerable body of literature relating to intermodal rail freight network design and operating concepts (see, e.g. Kreutzberger, 2008; Woxenius, 2007). Figure 3.4 presents a range of theoretical examples of the design of intermodal rail networks. In the Begin-and-End (BE) network, direct services are provided between each "begin terminal" and "end terminal". This requires either substantial flow volumes to make regular (e.g. daily) dedicated origin–destination provisions viable or, where volumes are low, relatively infrequent services. The other network variants all involve some sort of aggregation of flows, with the use of hub-and-spoke networks and/or trunk trains with feeder services. In practice, network operations often do not conform precisely to any of these networks, as the following British case study demonstrates.

Case study: Intermodal markets and network strategies of the British rail freight operators

Intermodal rail freight in Britain is dominated by maritime flows serving the key deep-sea container ports, although there are also active domestic and continental markets. The specifics are dynamic as routes are introduced/withdrawn or customer contracts change hands, but Table 3.2 summarises the situation in January 2015. Maritime intermodal accounted for 83% of intermodal trains running at that time. The majority of the remainder were domestic, with just 4% of the total being continental.

Competition between Freight Operating Companies (FOCs) has developed since rail privatisation in the mid-1990s. The FOCs involved in British intermodal rail freight have adopted different network strategies. Both Freightliner and DB Cargo emerged from the nationalised British rail when the rail system was privatised in the mid-1990s, while DRS and GB Railfreight entered the market after this time. The former have wider-ranging involvement in intermodal activities and more varied network strategies than the latter, reflecting broader European experience where new entrants typically have a narrower focus. Freightliner was formerly British rail's container division, with a monopoly on intermodal activity

Table 3.2 British intermodal rail freight markets (excl. niche ones), by FOC (as at January 2015)

Intermodal Market	*DB Cargo*	*DRS*	*Freightliner*	*GB Railfreight*
Maritime:				
Felixstowe	X		X	X
Southampton	X		X	
London Gateway	X		X	
Tilbury			X	
Domestic	X	X		
Continental (Channel Tunnel)	X			X
Indicative no. of trains per week	118	82	551	74
Indicative no. of terminals served	13	8	17	8

Source: Allan Woodburn.

at the time of privatisation; DB Cargo (as English, Welsh & Scottish Railway) started out with entirely non-intermodal traffic, later adding intermodal services to its portfolio. More historical background can be found in Woodburn (2012).

Of the four FOCs, Freightliner has the most extensive network coverage: it dominates the maritime market, with the vast majority of its services being direct trainloads between ports and inland terminals and vice versa. This most closely resembles the BE network in Figure 3.4, but not all begin terminals and end terminals in the network are connected with each other. Freightliner also operates a small number of block wagonload services serving the ports from a hub yard in Crewe, with spokes to terminals in the northwest of England, essentially a variant of the hub-and-spoke network shown in the diagram. These terminals receive a combination of direct trainloads and portions from the ports. The other three FOCs exclusively operate direct trainload services between terminal pairs, either port–inland or inland–inland. These mostly operate on a straightforward out-and-back basis between terminal pairs, but there are some examples of the interworking of locomotives and wagon sets between different routes to make more efficient use of resources. These flows from the other operators most closely resemble the BE network.

RAIL NETWORK INFRASTRUCTURE AND TRAIN OPERATING CHARACTERISTICS FOR INTERMODAL FREIGHT

In this section, the nature of intermodal rail freight operations is considered from the interrelated network infrastructure and train operations perspectives. First, the physical network infrastructure characteristics are discussed, followed by the key issues relating to train scheduling and operations.

Network infrastructure

Rail infrastructure is made up of the permanent way (i.e. tracks), providing a guided path for the passage of trains. Switches and crossings are provided to allow trains to change from one track to another, and to give access to and egress from rail freight terminals and yards. Signalling systems, mostly based on the subdivision of routes into block sections, control access to the infrastructure and ensure safe operation by maintaining the physical separation of trains on the network. The nature of the permanent way and signalling system influences the capacity and flexibility of the rail network, including the numbers of trains that can run, their operating speeds and their schedules.

The track–train interface is important in ensuring safe and efficient rail operations. Infrastructure managers are typically responsible for setting the rules that govern the use of their infrastructure, rules which are based on network capability. As an example, Network Rail (the British infrastructure manager) identifies eight variables influencing rail network capability for freight: track and route mileage, electrified track miles, permissible line speeds, gauge capability, route availability, length capability, gradients and total tonnage capability (Network Rail, n.d.). Variables with the most influence on intermodal flows are discussed briefly here and their impacts on train paths and operating speeds are dealt with in the next section. Of particular importance to intermodal rail freight are

- Gauge capability: 'The ability to move a railway vehicle and its load on a particular part of the network depends on the height and width profile, known as loading gauge, of the route concerned' (Network Rail, n.d.). Essentially, the combined wagon and unit load dimensions must fit within the loading gauge of the route to be used. Standard reference profiles have been developed by the International Union of Railways (UIC) for a range of gauges, taking vehicles and structures into account, and within the EU, some aspects of gauge capability are mandated by technical specifications for interoperability (TSI) (RSSB, 2009).
- Train length capability: Maximum train length is governed by the length of passing loops, yard and terminal sidings and, in some cases, the characteristics of the signalling system. For lighter-weight trains, often including intermodal freight, length rather than weight is often the constraint on how much can be carried per train. In Europe, key corridors are being upgraded for 775 m trains, but there are often infrastructure constraints limiting train lengths to around 400–600 m (i.e. 60–90 TEU per train).
- Route availability: Fundamentally, this refers to the maximum axle weight permissible on a given route. Despite intermodal services often being relatively lightweight, the combination of wagon design and unit load needs to be considered to ensure that axle loadings remain within safe limits and do not cause damage to bridges and earthworks. Maximum axle loads are highly dependent on the design and maintenance of the infrastructure, but 20–25 t per axle is typical.

Loading gauge is a particular issue for intermodal rail freight, since unitised traffic often has larger dimensions than traditional rail wagons and, in particular, the top corners of unit loads can be problematic for clearance of bridges and tunnels. Occasionally, the width of intermodal units can also be a critical factor, particularly on passenger routes with stations with high platforms. In North America, many routes have a sufficiently generous loading gauge to allow double-stacking of containers on rail wagons but this is not possible within the European context. On the routes with the most generous gauge, this is sufficient to allow two 9'6" high containers to be double-stacked. The United Kingdom generally has a more constrained loading gauge than other European countries, but considerable efforts have been made to provide a strategic network of routes capable of carrying 9'6" high containers on standard wagons. There are pressures elsewhere to provide routes able to carry larger intermodal units. For example, as part of the European Rail Freight Corridor 2 (North Sea-Mediterranean), investments are being made to create a corridor through Switzerland and Northern Italy capable of transporting 4 m high semi-trailers on commonly used wagons (SBB, 2016). Information about loading gauge restrictions and requirements can generally be obtained from the infrastructure manager, for example from Network Rail (http://www.networkrail.co.uk/aspx/10547.aspx) in the United Kingdom.

Train operating characteristics

Dedicated intermodal services are typically some of the fastest freight trains operating on the rail network, certainly from a European perspective; maximum speeds of 120 km/h (75 mph) are common and higher speeds are sometimes possible. In most cases, intermodal services will need to be allocated a train path through the network from origin to destination, alongside all of the other trains using the network. Depending on the level of network usage, the timetabling exercise may lead to a compromise solution rather than the optimal train path. Where intermodal services are offered on a shared-user basis, the timetable is typically fixed and often publicised so that potential users are aware of the service offer. Dedicated services for a single client may operate more flexibly according to demand, but typically will still need an allocated train path unless there is no interaction with other rail traffic. On a mixed traffic rail network, where passenger and freight trains share the same infrastructure, the ability to operate at speeds more similar to passenger trains can help with timetabling. The faster the maximum speed, the more likely the train will be able to keep moving rather than be sidelined in a yard or passing loop to allow faster passenger trains to overtake, thus increasing the average speed and reducing the end-to-end journey time. Also, the typically lighter-weight nature of intermodal services, certainly compared with the bulk traffic of a similar train length, can offer train acceleration benefits which again can help with gaining a better train path. Typical average speeds are considerably lower than the maximum operating speed, with averages in the region of 50–80 km/h (30–50 mph) being common.

EQUIPMENT FOR INTERMODAL RAIL FREIGHT OPERATIONS

Intermodal freight trains generally conform to the norm of one or more locomotives hauling a set of wagons. There are some operations using alternative approaches (e.g. freight multiple units, which are self-propelling trains without the need for a separate locomotive), but these have not been established in anything other than niche markets. This section focuses on the two key types of rail equipment: locomotives and wagons.

Locomotives (and associated staffing)

The traction considerations for intermodal freight train operations are fairly straightforward. For mainline operation, the two main requirements are that the locomotive(s) provides sufficient haulage and speed capabilities for the route taken. As mentioned previously, intermodal services are often lighter in weight than traditional bulk services, although they regularly have payloads of 1000–2000 t in Europe, with heavier services in North America. Heavy haul freight locomotives often have a maximum speed of 100 km/h (62 mph) or less, so are not suitable for intermodal services scheduled for faster speeds. A brand new freight locomotive typically costs around €2–4 million, although this very much depends on its characteristics, the order quantity and any maintenance agreements. Locomotives are often leased from manufacturers or leasing companies rather than bought outright by rail freight operators.

Additional traction may be needed at (or near) intermodal terminals or at marshalling yards en route. The loading/unloading tracks within terminals, particularly older ones, are often not capable of handling a full-length train. Where trains need to be split into two or more portions, either at the terminal itself or using nearby buffer tracks, a shunting locomotive may be required if it is not practical to use the mainline locomotive. This may particularly be the case when electric traction is used for the mainline operation, since overhead line wiring poses challenges for handling equipment (e.g. gantry cranes, reach stackers) at terminals. It is therefore common to find diesel shunting locomotives at intermodal terminals to split/join train portions and to move wagons around the terminal for loading/unloading. They can also be useful for removing specific wagons from a set for planned maintenance or because of faults. In small terminals, train shunting is often carried out by the mainline locomotive.

For some network operating concepts, notably with block wagonloads and/or hub-and-spoke networks, train marshalling is required at locations en route to make up trainloads heading in a particular direction. There are fewer, if any, restrictions on the use of electric traction in marshalling yards, so shunting of train portions may be carried out by diesel or electric shunting locomotives or, indeed, by the mainline locomotive.

Depending on factors such as local regulations and trades union agreements, the mainline operation may be carried out with a single driver at any particular time. Driver changeover points are normally influenced by train crew locations and requirements for drivers' breaks. Where train splitting/joining takes place, there may be a need for additional drivers and ground staff, depending on the level of shunting and the prevailing work practices.

Rail wagons

The wagons used need to be compatible with the unit loads carried, particularly with regard to the dimensions and the stowage/fixing requirements of the intermodal unit type (e.g. container, swap body, semi-trailer). The wagons can be individual wagons or permanently/semi-permanently coupled pairs (or more), often with articulated bogies. The mix of different intermodal types (e.g. ISO containers, swap bodies, semi-trailers) and the use of different unit load lengths can be a challenge for wagon design. As discussed earlier in the chapter, unit

Figure 3.5 Example of a multi-purpose intermodal wagon. (From Allan Woodburn.)

load lengths generally vary from 20' to 45', with some that are even longer. To some extent, the international classification of goods wagons implemented by the UIC in the mid-1960s and the subsequent European legislation on the TSI have helped with standardisation, although wagons used for intermodal purposes do not always conform to international standards. Wagon design, particularly the usable platform length relative to the total wagon length and the extent to which the available platform length can be utilised by the unit loads being carried, can influence the financial viability of intermodal rail freight. This is discussed further in a later section. Rail freight operators often lease wagons from manufacturers or leasing companies, and the cost is highly dependent on wagon design, lease length and the number of wagons involved.

Some wagon types are suited to a specific type of unit load while others can cater for different types of intermodal equipment. Figure 3.5 shows an example of a wagon which is able to carry ISO containers, swap bodies and semi-trailers. As shown, the wheels of the semi-trailer sit within the wagon, not far above rail height, to fit within the loading gauge. At intervals along the wagon body, hinged spigots can be seen. When carrying containers and swap bodies, the appropriate spigots can be placed in the upright position to fix the unit loads placed on them. Intermodal wagons typically have spigots positioned at appropriate locations along the wagon length to offer flexibility in the lengths of unit loads that can be carried. Locking pins can be used to secure the units if this is deemed necessary.

As discussed previously, the infrastructure loading gauge can be a limiting factor for the viability of intermodal rail freight. Where the gauge is too constrained for the unit loads to be carried on standard platform-height wagons, alternative wagon designs can be used to allow the combination of wagon and unit

Figure 3.6 Examples of different container wagon types. (From Allan Woodburn.)

load to fit. In the United Kingdom, the loading gauge is often too constrained to allow the conveyance of 9'6" high containers on standard platform-height wagons. Figure 3.6 shows three different types of wagon, coupled together on a maritime service. To the right of the picture is a standard platform-height wagon, with one 40' and one 20' container on a 60' platform length; this is a fairly common arrangement across Europe. The other two wagon types, a pocket wagon (centre) and a low-floor wagon pair (left), cater for higher containers within the same loading gauge. This is clear from the picture, since the 9'6" high containers carried by the pocket and low-floor wagons are no higher than the 8'6" high containers carried on the standard wagon. On some routes, there may also be constraints on the width of load that can be carried.

While most unaccompanied intermodal units need lifting on and off the wagons at terminals, there are some wagon designs that remove this requirement, allowing horizontal transfer instead. These are currently very much in the minority, but may offer a lower-cost solution for rail terminals where unit throughput is low. For example, the LOHR (2016) wagon conforms to UIC standards, can be used in a traditional way with vertical unit load transfer at terminals, but also offers a horizontal transfer option. A similar system has been developed by CargoBeamer (2015).

ECONOMICS OF INTERMODAL RAIL FREIGHT

Unlike many of the bulk traffics (e.g. coal, steel, construction materials), where rail regularly has a considerable cost advantage, intermodal rail freight often struggles to be cost-competitive with road haulage. The actual movement cost by rail is often cheaper than by road, but the door-to-door comparison between rail and road may be less favourable to rail because of terminal handling costs, feeder road costs and so on. For the rail operation itself, the greater the train payload (in weight and/or volumetric terms) and distance and the simpler the method of operation, the more favourable the economics are for rail movement.

Rail freight generally exhibits high fixed costs and low operating costs when compared with road haulage, though how this manifests is somewhat dependent on the nature of the cost structure applied to the rail industry. While it can be possible to identify approximate key rail freight costs (e.g. locomotives, wagons, labour, fuel/electricity), it is more difficult to ascertain the pricing policies because of limited transparency and commercial confidentiality concerns. The economics of intermodal rail freight are to a considerable extent dependent on the specific flow characteristics, and are influenced by the ways in which the costs, revenues and risks are allocated between the companies involved (e.g. rail freight operators, logistics service providers, shipping lines, freight forwarders, customers). Given the high fixed costs, the economics are improved when customer requirements are predictable and regular, since this smooths the utilisation of expensive assets. Recognising the challenges in operating viable intermodal rail freight flows, public funding may be available to assist with the costs of providing assets such as wagons or terminal equipment, or with train operating costs. Public support may come less directly in the form of improvements to rail infrastructure capacity and capability (see section on network infrastructure), or intervention in the operating costs and regulations for road haulage.

Time utilisation

Unless dedicated to intermodal traffic, most railway assets (e.g. locomotives, staff) can interwork with other types of rail freight, in theory at least. By contrast, by virtue of their design, wagons tend to be specific to intermodal services. In general, because of the relatively high costs of assets, particularly of locomotives but also wagons, there is considerable pressure to maximise their time utilisation. In the UK maritime market, GB Railfreight has built up its presence by focusing on routes where a round trip (including terminal time) can be carried out within 24 hours, leading to a higher-than-average time utilisation of locomotives and wagons. From original survey work, it has been found that it is not uncommon for the same locomotive and wagon set to spend an entire week on the same out-and-back daily diagram between Felixstowe and a particular inland terminal. Freightliner, with its more extensive and complex network, has some straightforward out-and-back workings, but also much interworking of wagon sets between different routes and, indeed, the splitting and joining of wagon sets within terminals or at the hub yard en route.

On-train capacity utilisation

An investigation of the British maritime intermodal rail freight market, based on comparable original observation surveys in 2007 and 2015, found that the average train carried 25% more TEU by 2015 (see Table 3.3). This improvement

Table 3.3 Average TEU load per UK maritime intermodal train, by port (2007–2015)

Port	*2007 Survey*	*2015 Survey*	*% Change*
Felixstowe	50.73	57.15	12.7
London Gateway	–	41.23	n.a.
Southampton	38.64	55.40	43.4
Thamesport	45.85	–	n.a.
Tilbury	29.86	41.78	39.9
Total	43.76	54.68	25.0

Source: Author's surveys.

resulted from a combination of an increase of 17% in the average on-train capacity and 8% growth in the average utilisation of train capacity (i.e. TEU carried as a percentage of TEU capacity). Put another way, the estimated number of TEU carried per annum rose from 1.26 million in 2007 to 1.58 million in 2015 with the same level of train service provision in terms of the number of port departures/arrivals per week. Likely reasons for this improvement in efficiency include

- Increasing intra-rail competition in the maritime intermodal market
- Loading gauge enhancements on key corridors allowing the greater use of more space-efficient standard platform-height wagons
- Other infrastructure improvements providing the capability for the operation of longer trains
- Investment in modern, more powerful locomotives and new wagon designs better suited to carrying the increasingly dominant 40' containers in the deep-sea maritime market

There is a considerable challenge in matching wagon types to unit loads to achieve high levels of on-train capacity utilisation. Wagon platform lengths of 60' (i.e. 3 TEU) are common for maritime flows (see the example in Figure 3.6), dating from the time when the proportions of 20' and 40' containers were fairly similar. With the reduction in the proportion of 20' containers in deep-sea shipping, it is more difficult to fully load these wagons. As a consequence, new wagon designs with 40', 45' or 80' platform lengths are becoming more common across Europe. The challenges of fully utilising the available wagon platform length are also felt in the continental/domestic market, where unit load sizes vary considerably, and, most particularly, where operators try to combine maritime and continental/domestic traffic on the same train. Either a compromise has to be made, whereby wagons are sufficiently flexible in their design to cater for the spectrum of unit loads but with a trade-off in utilisation levels, or different wagon designs are needed to be able to efficiently cater for each of the unit load types, with implications for overall fleet planning and utilisation. Wagon loading may also be influenced by axle load limits, meaning that wagons may not always be able to be fully loaded.

Figure 3.7 Example of an accompanied intermodal rail freight service. (From Allan Woodburn.)

ACCOMPANIED INTERMODAL RAIL FREIGHT

Accompanied intermodal rail freight (often referred to as rolling road, rolling highway, Rollende Landstraße [RoLa] or similar) is substantially different from the unaccompanied operations that have been the main focus of this chapter. Accompanied provision makes up less than 10% of the intermodal rail market in Europe, but it is important to highlight its key characteristics. This type of activity tends to manifest itself as discrete shuttle operations of fixed formation wagon sets, generally in locations where there are physical barriers preventing (or limiting) the use of road haulage. In addition to the wagons carrying the road vehicles, a passenger coach is normally provided for the lorry drivers to travel in. The two main examples of accompanied services in Europe are in the Alpine region, transiting Austria and Switzerland, and through the Channel Tunnel from France to the United Kingdom. Figure 3.7 shows a typical example of an accompanied intermodal service on the standard rail network, in this case crossing the Austrian Alps. The passenger coach for the lorry drivers is visible immediately behind the locomotive, followed by low platform wagons carrying the lorries.

There is a significant weight penalty associated with the carriage of entire road vehicles, with impacts on rail wagon axle loadings. There may also be loading gauge challenges because the overall height of the load is greater when the entire vehicle is carried. Unless the infrastructure is designed with the carriage of full vehicles in mind, as was the Channel Tunnel, accompanied intermodal transport requires very low platform wagons with small wheels in order to fit within the loading gauge. As a consequence, this form of intermodal activity tends to result in higher costs, both for the wagons and the infrastructure. Furthermore, standard intermodal terminals designed for the vertical transfer of unit loads are incompatible with accompanied intermodal transport, where ramps are required to allow road vehicles to be driven on/off the rail wagons. In some cases, terminal facilities and costs can be shared with passenger vehicle shuttles (e.g. Channel Tunnel).

Not surprisingly, the economics of accompanied intermodal transport are challenging and often rely on government support (recognising the environmental benefits of using rail in sensitive areas) and/or restrictions on road haulage activity. While accompanied intermodal transport is typically more expensive to the user than the unaccompanied provision, both in terms of the transport cost itself and in additional labour costs for the vehicle's accompanying driver, it can be an attractive option where alternatives are limited or are equally costly financially or time-wise. Depending on the duration of the rail transit, it may be possible for the lorry driver to take his/her statutory break or rest period while the lorry keeps moving towards its destination. Conveyance of the entire road vehicle can also simplify the overall operation since the vehicle stays together at all times so there is no need to position road tractor units at terminals.

FUTURE DIRECTION: OPPORTUNITIES TO IMPROVE INTERMODAL RAIL FREIGHT OPERATIONS

A number of practical developments have been made that should improve the viability and performance of intermodal rail freight. Across Europe, the implementation of rail freight corridors (RFCs) (OJEU, 2010) should provide greater capacity on core routes, improve service quality and allow longer, heavier trains to operate. The promotion of intermodality is a key element of the RFCs, with terminals and route infrastructure considered together as part of each corridor. Infrastructure improvements to allow 775 m long trains as standard should be particularly beneficial to intermodal flows, as train operating costs should rise at a slower rate than the increase in on-train capacity. Unit load costs by rail should therefore decrease, helping to make rail more financially viable. In combination with a focus on service quality (e.g. better train paths to allow faster journey times), intermodal rail should better meet customer requirements.

Technological developments within the rail industry are also important. Innovative wagons allowing the simpler and faster terminal transfer of unit loads were mentioned previously. Another contemporary development is the production of dual-mode locomotives by major manufacturers, including Bombardier, Siemens and Vossloh (now Stadler). These are electric locomotives, ideal for long-distance intermodal services, which have 'last-mile' diesel capability for branch lines into terminals, and for shunting trains in yards and terminals where lines are often not electrified. The use of a single locomotive design capable of electric performance on the main line and diesel power for the 'last-mile' is another opportunity for increased operational flexibility and lower operating costs.

Rail freight operators are often seen by customers as not being particularly responsive to their transport needs (see, e.g. ORR, 2012; Directorate General for Internal Policies, 2015). This can be a particular issue for unitised goods flows, since they are often more time-critical and of higher value than other rail freight markets. There is a role for logistics service providers (LSPs) to act as intermediaries between rail freight operators and customers, aggregating volume for viable rail operation and demonstrating an understanding both of the customer requirements and the characteristics of rail freight operation. For example, LSPs (e.g. WH Malcolm, The Russell Group, Stobart Rail) have been instrumental in the growth of domestic intermodal services in the United Kingdom, working with major retailers such as Tesco and Marks & Spencer (FTA, 2012).

CONCLUSION

This chapter has introduced the reader to key aspects of the operation of intermodal rail freight services, covering issues such as key markets, network design, typical equipment used and economic aspects. A viable intermodal rail freight provision is often challenging, since there is strong competition from the road-only alternative, but it is an increasingly important sector of the rail freight market in many countries. As identified towards the end of this chapter, there is considerable political support for intermodal rail freight at the European level and much of the planned rail network investment is likely to benefit intermodal activity to a greater degree than other types of rail freight.

REFERENCES

CargoBeamer. (2015). CargoBeamer: Environmentally friendly rail transport for all semi-trailer, Available at http://www.cargobeamer.eu/Flyer-CargoBeamer-Alpin-2016-852270.pdf (accessed 7 May 2016).

Council of the European Union. (1992). Council Directive 92/106/EEC of 7 December 1992 on the establishment of common rules for certain types of combined transport of goods between Member States. *Official Journal of the European Communities.* 368: 38–42.

Directorate General for Internal Policies. (2015). Freight on road: Why EU shippers prefer truck to train. Produced by Steer Davies Gleave at the request of the European Parliament's Committee on Transport and Tourism, IP/B/TRAN/FWC/2010-006/LOT1/C1/SC10. European Union. Brussels.

European Commission. (2015). Analysis of the EU Combined Transport: Final Report. Prepared by KombiConsult GmbH, Intermodality Ltd., PLANCO Consulting GmbH and Gruppo CLAS S.p.A. European Commission. Brussels.

FTA. (2012). *On track! Retailers Using Rail Freight to Make Cost and Carbon Savings.* Freight Transport Association (FTA). Tunbridge Wells.

Kreutzberger, E. (2008). The innovation of intermodal rail freight bundling networks in Europe: Concepts, developments, performances. PhD thesis. Delft University of Technology.

LOHR. (2016). The LOHR UIC wagons. Available at http://lohr.fr/lohr-railway-system-en/the-lohr-uic-wagons/ (accessed 7 May 2016).

Malcolm Group. (2014). The big box has arrived.......and it's a game changer. Available at http://www.malcolmgroup.co.uk/news/ (accessed 7 May 2016).

Network Rail. (n.d.). Network facts. Available at http://www.networkrail.co.uk/aspx/10536.aspx (accessed 11 April 2016).

OJEU. (2010). Regulation (EU) No. 913/2010 of the European Parliament and of the Council of 22 September 2010 concerning a European rail network for competitive freight. *Official Journal of the European Union (OJEU)*. 276: 22–32.

ORR. (2012). ORR Freight Customer Survey Report 2012. Produced on behalf of ORR by AECOM. Office of Rail Regulation (ORR). London.

ORR. (2016). Data portal: Freight rail usage. Available at https://dataportal.orr.gov.uk/browsereports/13 (accessed10 October 2016).

RSSB. (2009). Guidance on gauging, Rail Industry Guidance Note GE/GN8573. Rail Safety and Standards Board (RSSB). London.

SBB. (2016). The new Gotthard Tunnel: Switzerland through and through. SBB AG. Bern.

UIRR. (2015). UIRR Report: European Road-Rail Combined Transport 2014–15. International Union for Road-Rail Combined Transport (UIRR). Brussels.

Woodburn, A. (2012). Intermodal rail freight activity in Britain: Where has the growth come from? *Research in Transportation Business and Management*. 5: 16–26.

Woxenius, J. (2007). Generic framework for transport network designs: Applications and treatment in intermodal freight transport literature. *Transport Reviews*. 27 (6): 733–749.

Chapter 4

Inland waterway operations

Bart Wiegmans and Ron Van Duin

INTRODUCTION

In intermodal freight transport and logistics, inland waterway (IWW) operations also play an important role. In this chapter, IWW operations are classified into four related categories: (1) infrastructure; (2) transport, port and terminal operations; (3) management operations; and (4) challenges influencing operations.

INLAND WATERWAY INFRASTRUCTURE OPERATIONS

Infrastructure operations refer to operations involved in designing, building, financing and maintaining IWW infrastructure. IWW infrastructure consists of rivers, canals, locks, bridges and intermodal terminals (both inland and deep-sea), as shown in Table 4.1.

In general, the design and construction of rivers are not necessary as they already exist in nature. River maintenance is a crucial activity that is financed by public sources and managed by the Ministry of Transport. Canals, locks and bridges are also funded and managed in the same way. In addition, capacity problems can emerge for canals, locks and bridges. For instance, due to vessel enlargements and traffic growth, canals might become too narrow (see Figure 4.1).

A lock is part of the infrastructure network and the skipper often needs to slow down (or sometimes wait) to pass a lock. The lock separates waterways due to a difference in water level or salinity (saltwater/freshwater). A lock can consist of one or more chambers with certain dimensions (length, width and allowed vessel depth). Several challenges exist for locks: the mix of freight and recreational water travel leads to safety issues; a lock might be tide dependent (loaded deep draught vessels can only enter a lock during high tide); and traffic and vessel size growth might lead to increasing waiting times at locks (see Figure 4.2 for an example of a lock).

Thus, capacity issues can also arise for locks now and then (Smith et al., 2009). These capacity issues can sometimes lead to design and building operations that aim to create additional canal, lock and bridge capacity. Besides the maintenance costs, additional financing is needed to increase capacity. Usually, these investments are quite large and difficult to realise due to a lack of political support. In terms of

Table 4.1 Relations between infrastructure elements and operations

	Design	*Build*	*Finance*	*Maintain*
Rivers	No	No	Yes	Yes
Canals	Sometimes	Sometimes	Yes	Yes
Locks	Sometimes	Sometimes	Yes	Yes
Bridges	Sometimes	Sometimes	Yes	Yes
Terminals	Sometimes	Sometimes	Yes	Yes

Figure 4.1 A canal used for IWW. (From Container Terminal Alpherium.)

design, construction, finance and maintenance, terminal infrastructure is comparable with canals, locks and bridges. Terminals require regular finance and maintenance, while the design and construction of new terminals or the extension of existing ones is only needed now and then. The main difference is that terminals are not owned by the Ministry of Transport, but by private operators. Together, all these infrastructure elements form an IWW network (see Figure 4.3).

Vessels are needed to operate inland waterway transport (IWT) services. There is a wide variety of IWW vessels, ranging from very small units with a carrying

Figure 4.2 Beatrix lock and the Lekkanaal. (From https://beeldbank.rws.nl/MediaObject/Details/Scheepvaard%20in%20de%20sluiskolk_367046?resultType=Search&resultList=367061,367060,367059,367056,367055,367046,359679,359678,359677,359676,349046,349045,349044,349043,349042,349041,348701,345981,345980,345945.)

capacity of just a few tonnes (usually self-propelled vessels) to large push tows propelling many barges with a carrying capacity of several thousand tonnes (see Table 4.2). Self-propelled vessels can be operated by a small crew, are streamlined and easy to build. Push tows are more expensive to build as they require a separate pusher and several cargo units (barges). The pusher and the barges are connected by steel wire ropes and the connection of the pusher with the barges can be a time-consuming (and thus labour-intensive) process. The large push tow combinations can realise scale economies (in loading capacity and fuel consumption) and the barges can be used as floating stock (see Figure 4.4).

Aside from vessel size, the size of the company that owns and operates the vessels is important. In the United States, about half of the market is occupied by five large operators. In Europe, the majority of IWT companies are small and own just one to three vessels. In China, there is a combination of these two models with many small companies along with some very large IWT operators. Next to the size of the respective vessels, the sector in which the vessels operate determines their operational characteristics.

In general, the majority of freight flows transported by IWW are dry bulk. The second most important sector is liquid bulk (such as oil, chemicals and refined products) and these products are often transported in tank vessels. A relatively small but rising sector is containers. In general, most vessels are designed in the country within which they are going to operate. The building of the vessels is often executed in low-cost countries (such as China or Eastern European countries). The maintenance of the vessels takes place in the country of operation. Financing the vessels is one of the most critical points as these vessels can cost

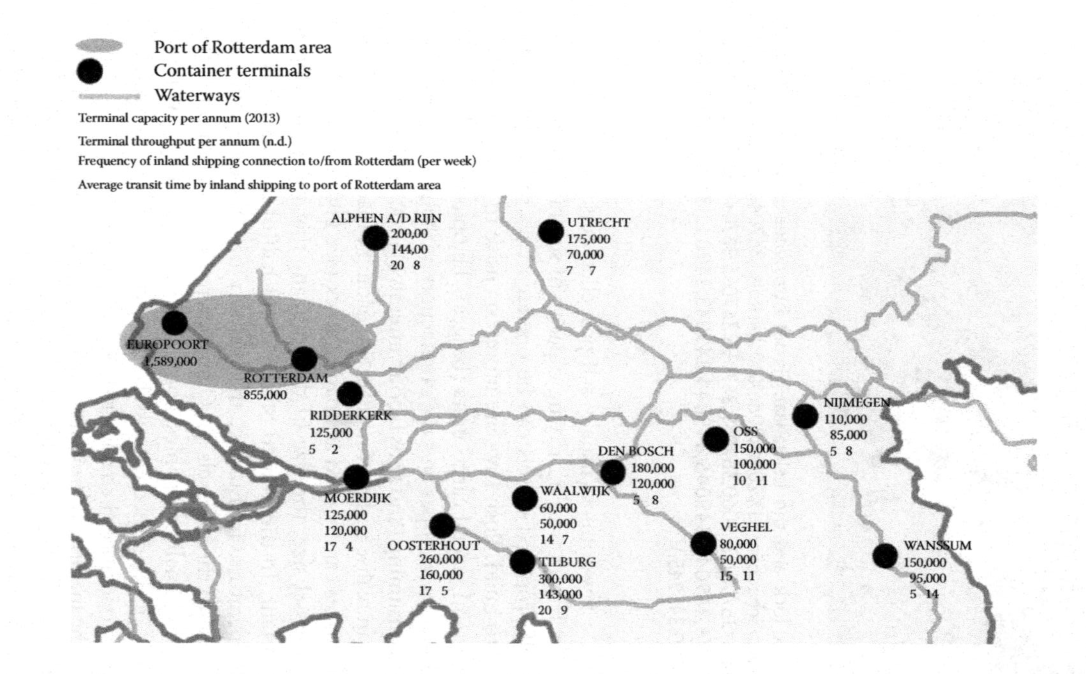

Figure 4.3 IWW network for the Port of Rotterdam. (From Fan, Y., The design of a synchromodal transport system: Applying synchromodality to improve the performance of current intermodal freight transport system, Master's thesis, Delft University of Technology, 2013.)

Table 4.2 Fleet statistics: Number of vessels

	W-Europe	*Europe Danube*	*United States*	*China*
Self-propelled				
Dry cargo	6.753	373	635	n.a.
Tank	1.992	37	2	n.a.
Total#	8.745	410	635	132.000
Push barges				
Dry cargo	3117	2559	23418	n.a.
Tank	155	233	3220	n.a.
Total#	3.272	2.792	26.638	33.000
Pushers	1039	422	3442	n.a.
Total	n.a.	n.a.	n.a.	165.000

Source: Adapted from Hekkenberg, R. and Liu, J., *Inland Waterway Transport: Challenges and Prospects*, Routledge, London, 2016.

Figure 4.4 A tugboat pushing barges up the Monongahela. (From https://rutheh.com/2010/03/10/tugboat-pushing-barges-up-the-monongahela.)

millions of euros to construct. This leads to high fixed cost levels and attaches greater importance to capacity filling (high utilisation). Currently, small growing sectors in IWT are palletised goods, the transport of flour and vehicles, and freight delivery by city canals, especially construction materials and garbage collection.

All IWW infrastructure is built to cater for all different freight categories transported by IWWs. Countries where IWT is important are Europe, China and to a lesser extent the United States (see Figure 4.5 for major IWW infrastructures). In the next section, we will discuss the inland waterway transport and terminal operations in more detail for the three most important freight categories: dry

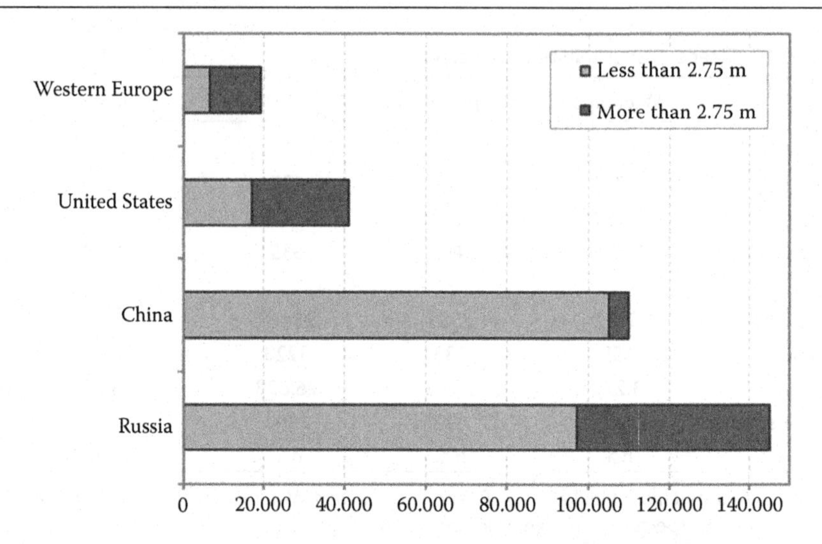

Figure 4.5 Length of major inland waterway systems, 2000. (From Jean-Paul Rodrigue, Dept. of Global Studies & Geography, Hofstra University, 2012.)

bulk (such as molten solids, powders and small granules, delicate bulks, e.g. root vegetables, and other solids), liquid bulk (gases in a gaseous state, liquefied gases, gases in solution, volatile liquids and other liquids) and containers. Not included here are less important segments such as roll-on roll-off and other general cargo.

INLAND WATERWAY TRANSPORT AND TERMINAL OPERATIONS

Dry bulk: Transport and terminal operations

One of the major tendencies in the IWW dry bulk transport market is that of overcapacity (van Hassel, 2015). Vessel capacity is often added in large volumes, which leads to overcapacity and thus cut-throat competition and low profits (or losses) for the skippers.

In general, dry bulk transport and terminal operations can be characterised by regular transport flows and thus smooth operations. The export of dry bulk flows often begins in the hinterland where dry bulk (e.g. coal, ores, grain) is loaded into a vessel and transported to a deep-sea port. The ports in the hinterland are often close to mines where iron ore, coal, limestone and other industrial raw materials are produced. In the deep-sea port, the dry bulk is stored at the deep-sea dry bulk terminal in order to collect as many IWW vessel loads as are needed to fill a deep-sea ship. The import of dry bulk flows works the other way around. A deep-sea vessel with dry bulk arrives at a terminal where the cargo is unloaded and stored at the terminal. At regular intervals, IWW vessels are filled with dry bulk (e.g. coal, semi-manufactured goods such as steel products and construction materials such as sand and cement) and the load is transported further inland, for example, to a coal burning plant. These inland ports are also located close to factories that use the raw materials or semi-finished goods that are produced overseas and imported via the deep-sea ports. Given the large scale of the actors involved in dry bulk flows, IWT is dominated by large corporations such as Arcelor-Mittal. These large firms often also control the production of raw materials, leading to large investments in their mines and heavy industry, thus lowering the importance of IWT operations in their total dry bulk supply chains.

Liquid bulk: Transport and terminal operations

Liquid bulk operations often begin in the deep-sea port. Liquid bulk loads consist mainly of oil and chemicals. Deep-sea ships often arrive with large liquid bulk volumes at large deep-sea ports. Their freight is unloaded and stored in the port area. In the port area, additional production activities are often executed in order to further upgrade liquid bulk products (e.g. oil is upgraded to gas, diesel, liquefied natural gas [LNG], etc.). After these production activities in deep-sea port areas, the liquid bulk is transported further inland by IWW to other production facilities (in the case of chemicals) or to storage facilities (in the case of gas, diesel and LNG). Vessel developments in the liquid bulk sector have been especially driven by the requirement for all newly constructed vessels to have double hulls in order to further improve safety. Also, in the liquid bulk IWW sector, a small number of large firms (e.g. Shell, Vopak) determine the origins and destinations of freight flows. These firms often produce the raw materials (oil) themselves and transport them to large factories in deep-sea ports where they can be upgraded to semi-finished or end-products that need to be transported further inland by IWT.

Containers: Transport and terminal operations

Transport and terminal operations in container transport are diverse in terms of origins, destinations and other variables. A general overview of all possible operations within a container transport chain is provided in Figure 4.6. If IWT is involved in such transport chains, road transport is then replaced by IWT.

In IWW container transport and handling operations, many different conceptual models to execute IWT operations exist. In Figure 4.7, an example of a conceptual model focused on deep-sea container flows that are further transported by IWW into the hinterland is depicted. In Figure 4.7, an IWW vessel sails to a deep-sea container port and visits different container terminals (the maritime container terminals and other container terminals) in the deep-sea port area in order to collect all containers destined for the inland container terminal (left side in Figure 4.7). The operations within the deep-sea port area are quite complicated and delays often cause IWW container vessels to spend up to one week in the port area collecting all containers destined for the hinterland. After all containers are collected in the port area, the IWW vessel sails into the hinterland

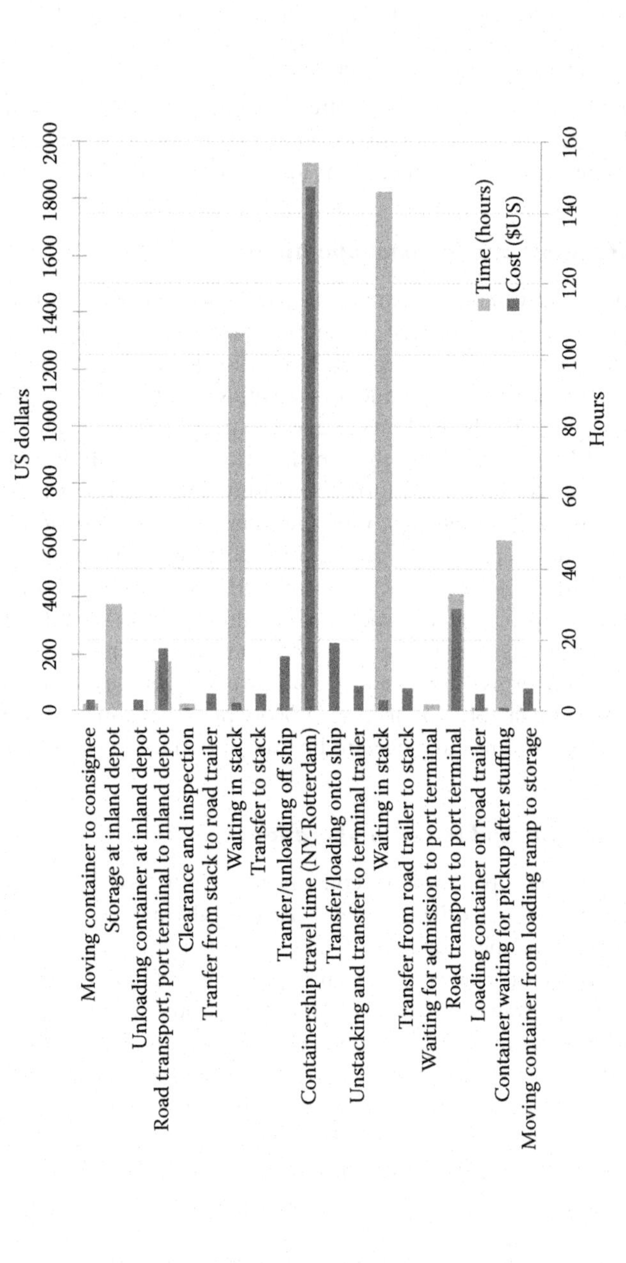

Figure 4.6 Overview of freight transport operations. (From Jean-Paul Rodrigue, Dept. of Global Studies & Geography, Hofstra University, 2012.)

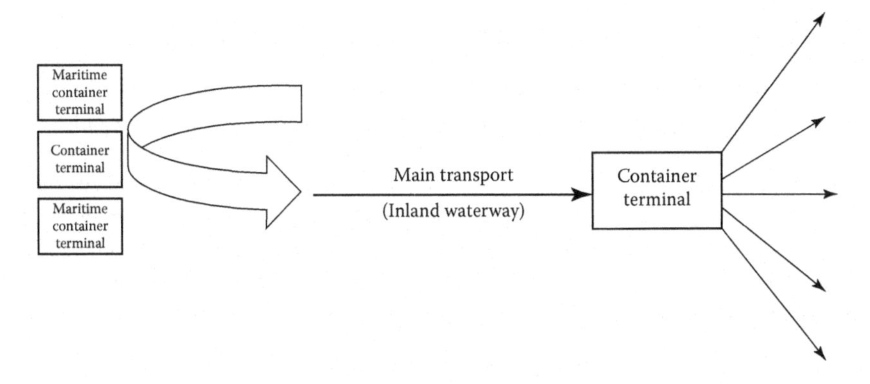

Figure 4.7 Containers transported by IWW to and from deep-sea ports. (Based on Wiegmans, B. and Konings, J.W., *The Asian Journal of Shipping and Logistics*, 31(2), 273–294, 2015.)

towards the destination terminal (main transport in Figure 4.7). On arrival at the container terminal, the end-haulage of the containers towards the final customer takes place (right side in Figure 4.7).

In Figure 4.8, an alternative conceptual IWT is depicted. This is an IWW palletised transport hub-and-spoke network with even more complicated transport and handling operations. Incoming IWW vessels and trucks need to be aligned in order to enable parallel exchange at the terminal (left side in Figure 4.8: the left terminal). In the terminal, pallets are bundled to ensure large enough IWT flows. Next, vessels (and sometimes trucks) sail to the next terminal (centre in Figure 4.8 between the pallet terminals). At the pallet terminal (the right terminal), the pallets are unloaded and end-haulage towards the final customer takes place.

Containers transported by IWW from deep-sea ports: Extended gate concept

An example of extended gate services is the extended gate concept of ECT. The 'extended gate concept' makes use of inland terminals (dry ports) as extensions of

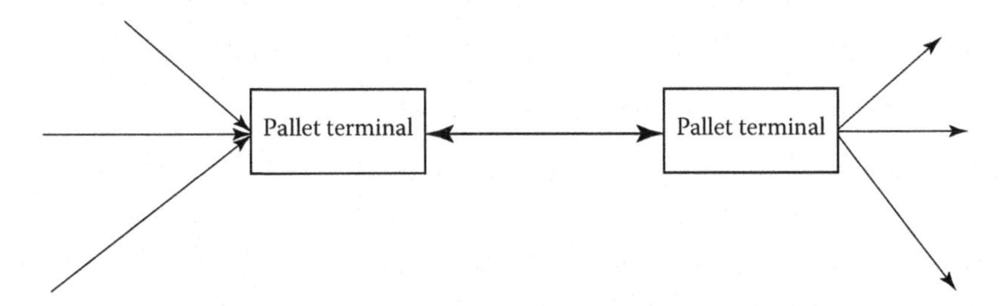

Figure 4.8 Overview of IWW palletised transport hub-and-spoke conceptual model. (Based on Mommens, K. and Macharis, C., *Journal of Transport Geography*, 34, 44–53, 2014.)

the main port of Rotterdam (see Figure 4.9). Different services (such as customs) are moved from the main terminals in Rotterdam further inland so that containers can be transported as fast as possible into the hinterland. The inland terminals in the EGS network are connected to the Port of Rotterdam by barge or rail connections. The 'extended gate', for example, means that customs are conducted at these inland terminals instead of at the Port of Rotterdam, decreasing the waiting times at the port and thus increasing throughput and revenues. The fact that rail and barge modalities are promoted in this concept also decreases road congestion around Rotterdam (Veenstra et al., 2012). The main focus in the EGS network is on barge services, but where possible rail services are also used.

Containers transported by IWW to deep-sea ports: Hub-and-spoke network

In Figure 4.10, an overview of an IWW container transport hub-and-spoke network is given. A hub-and-spoke system can be characterised by barges that would have previously been required to visit multiple terminals but now only need to visit one terminal (Konings et al., 2013). In the hub terminal, the containers on vessels coming from the hinterland to the port area (and also the container flows from the port area into the hinterland) are sorted according to their destination terminal. The advantages of a hub-and-spoke system include shorter vessel port turnaround times, more reliable barge services and higher crane and quay productivity for terminal operators due to the larger call sizes. However, sorting and exchange operations in the hub can incur additional costs and time. Handling in the hub also leads to extra handling when compared with the direct service network. In IWT, such hub-and-spoke networks are seldom found in practice. An interesting study has been carried out by Plompe (2014) into a hub-and-spoke system for the Port of Amsterdam, which acts as a hub connecting the Port of Rotterdam to the hinterland in the north of the Netherlands (Figure 4.10).

In his study, Plompe proved that if the intermodal inland waterway transport (IIWT) sector wants to utilise the hub-and-spoke network's scale advantages and potential for attracting new freight flows and activities, the costly and time-consuming transfer process of IIWT first needs to be improved. In theory, hub-and-spoke systems seem to offer several opportunities and advantages; however, in practice they prove difficult to implement while also continuing operations.

IWW palletised transport hub-and-spoke system

The transport of palletised goods via Belgian IWWs shows clear potential for palletised building materials and fast-moving consumer goods (Mommens and Macharis, 2014). Cost modelling shows clear economic potential for palletised IWT when both the producer and the customer are located close to an IWW. Comparable to IWW container transport, longer pre- and end-haulage distances will diminish the potential for palletised freight to be transported by IWW. In

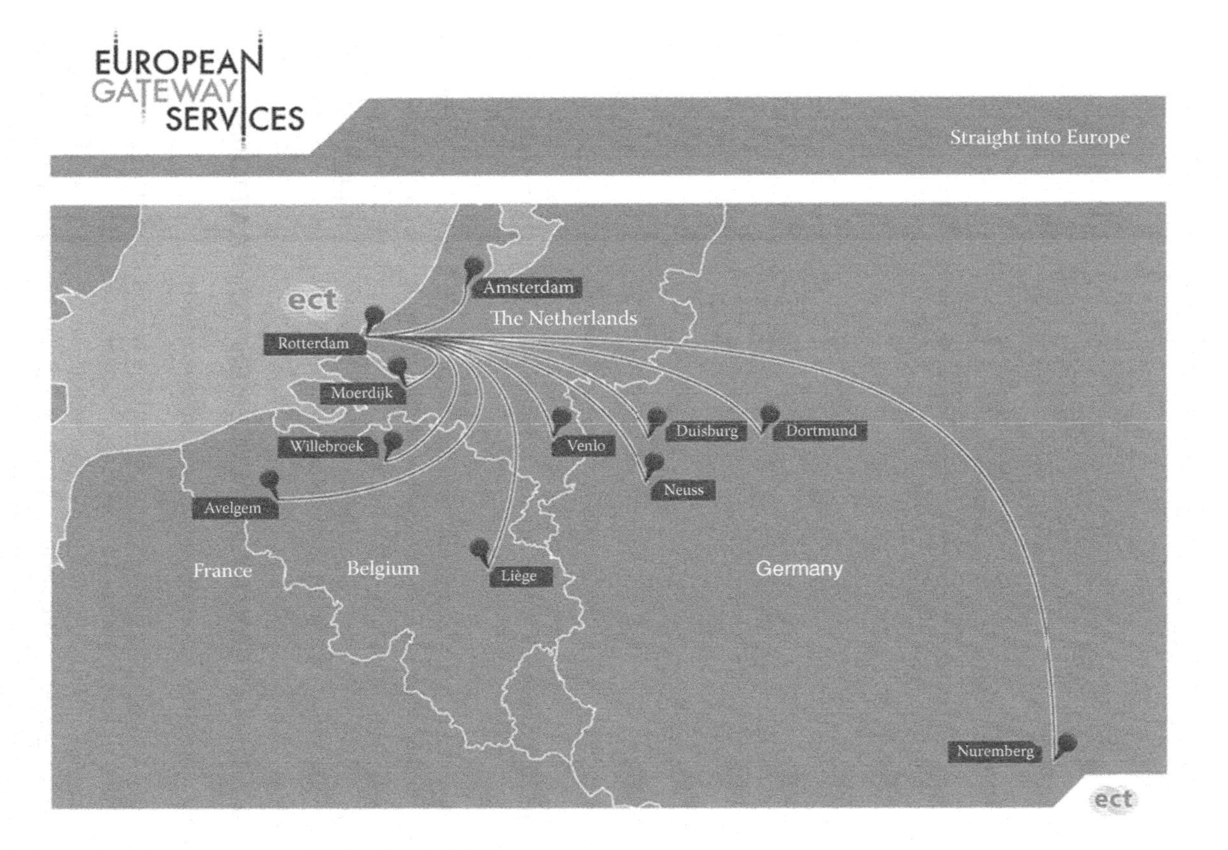

Figure 4.9 The extended gate concept of ECT. (From ECT.)

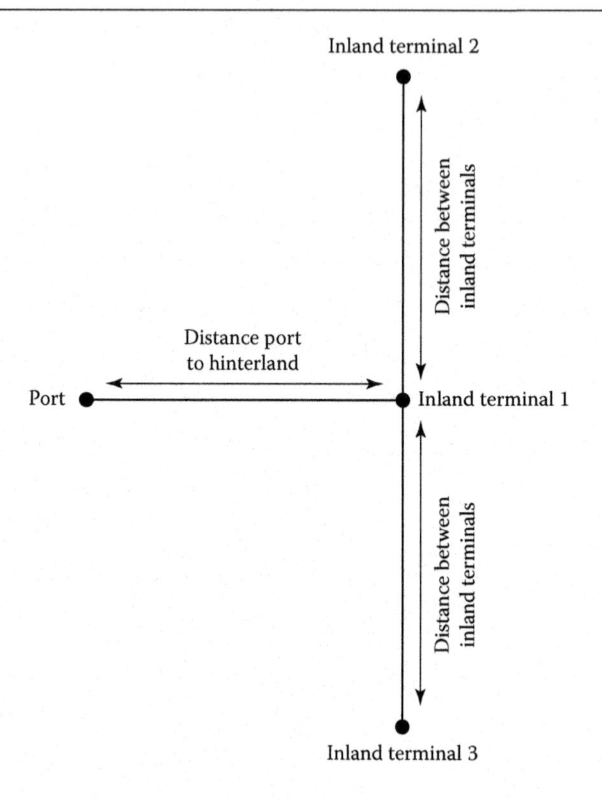

Figure 4.10 Example hub-and-spoke network for the Port of Amsterdam. (From Plompe, 2014.)

this respect, shorter pre- and end-haulage distances will lead to higher demands for a number of terminals to be implemented in the palletised goods network. In addition, environmental and societal benefits also result from the implementation of a palletised goods network. The initiative in Belgium is comparable with the failed initiative of Distrivaart in the Netherlands.

INLAND WATERWAY MANAGEMENT OPERATIONS

Inland waterway transport management operations

IWT processes require the involvement and interaction of numerous transport operators, such as shippers, IWT operators, skippers, road operators for pre- and end-haulage, inland port and terminal operators and logistics service providers (LSPs). In general, LSPs organise IIWT operations. Barge transport operators, terminal operators and pre- and end-haulage operators execute the actual transport and handling operations. For the barge transport operators, the most important operations are vessel operations, vessel utilisation and the role of the captain and crew. In IWT, different vessel sizes are in operation, leading to different requirements for the manning of the vessel. Furthermore, the type of operation – day operations (maximum 14 hours/day), semi-continuous (maximum 18 hours/day) or continuous operations (24 hours/day) – also influences the operation and logistical conditions. The loading degree of the barge and the resulting utilisation rate per time period are of importance when operating the barge. The utilisation rate in particular is influenced by the duration of the round trip of the vessel. Shorter round trips enable more round trips in the same time period and thus achieve a higher utilisation rate. In this case, the fixed cost can be shared among more load units and the average transport cost per load unit will decrease. The round trip time is further influenced by bridge characteristics (height determining vessel load), lock passage time and the terminal waiting and handling times. Incidentally, very high or very low water levels might also limit the loading capacities of the barges and vessels. Another operational characteristic of barge operations is scale economies, where the operation of larger barges can lead to a lower average cost per load unit (e.g. container). In the end, the IWT route and the terminal performance will determine the precise IWT logistics operations.

The working conditions on vessels depend on the vessels' flag, the relationship between the owner and the operator of the vessel and legislation (such as working times, annual leave, wages, pension schemes, disability benefits, etc.). It is increasingly difficult to man vessels operating in IWT due to ageing and unattractive prospects. IWT attractiveness can be improved through remuneration and social benefits, job flexibility (working hours, working time arrangements and time flexibility), job security, employee participation and skills development. However, the expected introduction of more information and communications technology (ICT) and semi- or fully automated vessels might reduce the demand for personnel. On the other hand, although vessels with semi- or fully automated operations and additional ICT features might have less crew on board, the requirements for the remaining crew might increase. Furthermore, new functions, such as the maintenance and repair of ICT and automated operations, might be introduced.

Transhipment operations: Terminals

At the IWW terminals, handling operations at the quay are executed by multipurpose or dedicated container cranes. For further handling and storage operations in the yard, mobile equipment (such as reach stackers) or more fixed equipment (rail-mounted cranes) are used. Different combinations of fixed and mobile equipment lead to different operation conditions at the terminal. In addition, the terminal has an office where planners and management are located. Next to the core handling and storage services of the terminal, other services (added value, container cleaning, end-haulage, etc.) can be offered. At IWT terminals, a wide variation of terminal configurations (i.e. layout), amounts and types of equipment and additional services can be found. Terminal operations can be further influenced by several conditions and limitations. For example, severe weather conditions might cause the temporary closure of the terminal while delays in IWT will also influence terminal operations and congestion in terminal handling might occur (e.g. the arrival of large IWW barges that must be unloaded or loaded quickly). Finally, noise and/or emission restrictions imposed by local governments might limit the terminal operating hours and thus influence its operations (van Duin and van der Heijden, 2012).

Inland waterway logistics management operations

Inside supply chains, logistics management operations generally consist of a physical flow and an information flow (e.g. invoicing, acknowledgement of receipt of payment). When customers consider using IWT, logistical decision makers inside the companies need information on several decision factors: (1) cost, (2) reliability, (3) flexibility, (4) door-to-door transit time and (5) information. Information in particular is increasingly important for logistics management operations. The aspects that logistics customers take into account when deciding whether to integrate IWT into their logistics chains vary depending on the freight type, the volume and so on. In Europe, in general, accessibility of the IWW network is not good. In addition, the consignee will often not have direct IWW access and thus end-haulage might be required. Several other logistics management operational developments place further pressure on the competitiveness of IWT. First, the customisation of products, e-commerce growth and other related developments in modern highly developed service economies result in a smaller amount of bulk cargo, while more end-products and general cargo (including containers) are transported. Secondly, current logistical and production concepts, such as *just in time/just in sequence* (JIT/JIS), are not in favour of the use of IWT. This has also led to smaller shipment sizes and more frequent shipments.

Several opportunities also exist for IWT to increase its role in logistics management operations. First, synchromodality is developing and this may enhance the competitive position of IWT. Synchromodality aims at the coordinated and improved usage of IWT, rail and road transport infrastructure. Synchromodality is a new concept in freight transport that has not yet garnered a precise definition. There is, however, general agreement that synchromodality encompasses an integrated view of planning and uses different transport modes to provide flexibility in handling transport demand (Behdani et al., 2016). If this results in the coordinated supply of transport services of different modes which are perceived by the customers as a single service, then synchromodality has been achieved. Secondly, new market segments for IWT could be developed, such as reverse logistics (waste flows) and biomass transport. The market segments that are currently important in IWT (dry bulk, liquid bulk and containers) will, in general, not demonstrate very high growth. An exception to this development is the market for LNG, which may be a promising market in Europe (IGU, 2015). So, if IWT wants to enlarge its market share and start growing faster than freight transport in general, then new market segments must be developed. However, increasing the relatively limited role of IWT in logistics transport operations might remain difficult.

INLAND WATERWAY CHALLENGES INFLUENCING OPERATIONS

When looking towards the future of IWW operations, five main challenges have been identified: (1) sustainability, (2) climate change, (3) cost and price, (4) truck platooning and (5) information exchange.

Sustainability challenge

In general (depending on the assumed variables), IWT is more sustainable in terms of carbon dioxide (CO_2) emissions in comparison with trucking and rail transport. However, truck transport is making enormous progress in terms of CO_2 exhaust emissions and in certain instances is already more sustainable than IWT. Furthermore, in terms of particulate matter 10 (PM10) and oxides of nitrogen (NOx), truck transport is already more sustainable than IWT. In order to meet the EU 20% targets, the transport sector is required to not only reduce the 1995 carbon footprint by 20% but also to compensate for the growth that has occurred since 1995. The EU facilitated the Green Efforts Project, a collaborative research project aimed at reducing energy consumption at terminals which commenced in 2012 and ended mid-2014 (Froese and Töter, 2014). This study illustrates how to understand the energy consumption profile of an IWW terminal. The calculation of energy consumption and the related carbon footprint by the terminal processes (and their equipment) has been well-documented in Geerlings and Van Duin (2011) and Van Duin and Geerlings (2012) (Table 4.3). The model shows the relevant factors that influence the consumption of fuel and electricity. Important factors for determining total energy consumption are the average distance per container movement, total container throughput and terminal

Table 4.3 Estimated emissions of CO_2 after the implementation of policies

Terminal	*Reference year (2006)*	*Compact terminal*		*Fast replacement of diesel equipment*		*30% blending of biofuel with diesel*	
	Emissions (kton/year)	*Emissions (kton/year)*	Δ%	*Emissions (kton/year)*	Δ%	*Emissions (kton/year)*	Δ%
Delta	71.30	23.81	−67%	55.47	−22%	57.35	−20%
Home	15.01	5.40	−64%	12.14	−19%	11.70	−22%
Hanno	1.20	0.26	−78%	1.03	−14%	0.94	−22%
APM	35.95	10.69	−70%	29.77	−17%	26.74	−26%
RST	10.76	5.99	−44%	9.79	−9%	9.25	−14%
Uniport	6.53	2.49	−62%	6.18	−5%	5.67	−13%
Total	**140.75**	**48.68**	**−65%**	**114.38**	**−19%**	**111.65**	**−21%**

Source: Geerlings, H. et al., *Journal of Cleaner Production*, 19(6–7), 657–666, 2011.

configuration. This study also shows that diesel-powered equipment is largely responsible for total CO_2 emissions at terminals.

These emissions are largely caused by the terminal layout, which influences the length of movements required for terminal equipment. New designs of current terminal layouts could reduce current emissions by up to 70%. The other suggested policy recommendation is to achieve a reduction in emissions by mixing currently used diesel with biofuels. Overall, this means that huge steps must be taken in order to further improve the sustainability of IWT and terminals and the competitive position of IWW. This will be difficult to achieve because if IWT has to comply with stricter emission regulations, vessels currently in use will have difficulties complying with stricter emission regulations. As a result, vessel engines will have to be either replaced (an investment leading to additional costs) or supported by catalysts (likewise causing additional investment). All in all, this will impose a large financial burden which the sector may struggle to bear.

Climate change challenge

A major challenge for IWW lies in climate change. In general, climate change leads to more frequent and longer periods of extreme weather conditions. Water levels can fluctuate significantly and in the case of low (or high) water levels, less cargo can be transported with inland ships due to drought restrictions. These fluctuations could increase in the future due to climate change, with wetter winters and dryer summers (Jonkeren et al., 2013). Low and high tides in particular can result in problems for IWT by reducing the loading capacity of vessels and barges. This reduction can account for 75% of the initial loading capacity, leaving a loading capacity of only 25%. An example of this problem occurred in 2015 when low water levels prevented traffic on the Rhine River. In the short term, customers might 'accept' such an event, but when it happens more often, the transport solution may be perceived as less reliable and customers may be encouraged to decide in favour of another transport mode.

Cost and price challenge

Service cost and price are important determinants of the competitiveness of IWT. Wiegmans and Konings (2015) showed that IWT in most cases is more than competitive with truck transport at distances of 600 km or more. In the medium distance category (200 km), they showed that – depending on the exact combination of variables – in most cases, IWT is lower in cost. In the short distance category (50 km), in all cases, truck transport is lower in cost than IWT. So, depending on the distance and the combination of variables influencing the cost for producing the IWT service, IWT can be competitive (see also Table 4.4 for the most important cost components of IWT).

Knowledge of the production cost of IWT services does not ensure knowledge of the price – it only gives an indication. The pricing of IWT services is practically

Table 4.4 Factor costs in inland waterway transport (reference date: 2008)

	Measure	*Rhine vessel (Class Va)*	*Rhine-Herne vessel (Class IV)*
Vessel characteristics			
Type of vessel		Motor dry freight vessel	Motor dry freight vessel
Capacity	TEU	208	90
Dimensions (L × W × D)	Metres	110 × 11.40 × 3.60	86 × 10.50 × 3.20
Tonnage	Tons	3.500	2.000
Fixed costs			
Capital costs	€/year	784.750	350.000
Labour costs			
Day operations	€/year	140.000	120.000
Semi-continuous operations	€/year	285.000	250.000
Continuous operations	€/year	660.000	510.000
Variable costs			
Fuel costs			
Loaded vessel	€/km	10	7.54
Empty vessel	€/km	4.78	3.62
Repair and maintenance costs	€/km	0.72	0.37
Overheads	€/year	n.a.	n.a.
Business hours			
Day operations	Hours/year	3.500	3.500
Semi-continuous operations	Hours/year	4.500	4.500
Continuous operations	Hours/year	7.800	7.800
Direct cost hour coefficient			
Day operations	€/hour	264	134
Semi-continuous operations	€/hour	238	133
Continuous operations	€/hour	185	110
Kilometre cost coefficient			
Loaded vessel	€/km	10.72	7.91
Empty vessel	€/km	5.50	3.99

Source: Adapted from NEA (2009).

a blank field. Market consultation indicates that relatively more expensive short IWW distances (in terms of cost) are cross-subsidised (in pricing) by relatively cheaper long distance IWT services. This means that the price of IWT services on short distances might be offered at a lower rate to the service production cost. The challenge for IWT is to maintain competitiveness with truck transport and continuously search for opportunities to reduce costs. For the sector as a whole, this may be quite difficult to achieve given the large proportion of fixed costs linked to the vessel and its financing.

Truck platooning challenge

In the future, a major challenge for the IWW sector might originate from the development of truck platooning. In truck platooning, several trucks might be automatically connected and when fully developed and operational, this may reduce fuel consumption by between 5% and 15%; the number of truck drivers might also be reduced to one (driver of the first truck) (Ashley, 2013). This will lead to increased sustainability and reduced costs for truck transportation if implemented. Truck platooning will also make truck transport even more attractive in terms of cost–price combinations. So, in order to stay competitive, at least equal efforts in terms of cost–price combinations are required from the IWT sector.

Information exchange

Information exchange and sharing is also becoming increasingly important for IWT and handling. The first system that was used in the 2000s was EDILAND, where relatively simple messages were exchanged between actors in the IWT chain (den Hengst-Bruggeling, 1999). Next, in major port areas such as Hamburg, Antwerp, Bremen and Rotterdam, cargocards were introduced in order to facilitate information exchange between all actors in the port area and also to regulate port and terminal access. A recent initiative that has been implemented Europe-wide is the river information system (RIS). RISs are mostly national traffic management systems enhancing swift electronic data transfer between water and shore through in-advance and real-time information exchange. An EU framework directive provides minimum requirements to enable cross-border compatibility of national systems. The RIS can support transport planning by providing IWW infrastructure-based information (such as water levels, vertical clearance information about road/rail bridges and lock operations). RIS also provides its users with position information on inland vessels and unpropelled barges as well as tracking services.

CONCLUSION

This chapter has discussed all the actual operations of the barge transport sector. From the discussion on infrastructure and vessels, it follows that different infrastructure aspects exist (such as rivers, canals, locks, bridges and terminals) which together form the inland waterway network (IWN). Given these infrastructures, four different activities (design, build, finance and maintain) can be distinguished. This can be further coupled with different parts in the supply chain, such as the deep-sea port, IWT, inland terminals and end-haulage, and with different freight transport segments. It is clear that all these variables lead to an overwhelming number of possible combinations of transport operations. However, central to operations are the transport operations (both main transport and begin- and end-haulage), terminal operations and logistics operations.

The main challenges that the IWW sector will face in the coming years are (1) sustainability, (2) climate change, (3) cost and price, (4) truck platooning and (5) information exchange. Sustainability developments such as electric trucks, climate change and truck platooning, will threaten the competitive position of IWT, especially when compared with trucking. Cost and price improvements and the possibilities of information exchange will have to make up for the threats expected to compromise the position of IWT, threats which indicate that huge effort is needed in the next few years by the IWT sector.

REFERENCES

Ashley, S. (2013). Truck platoon demo reveals 15% bump in fuel economy. SAE Off-Highway Engineering. Retrieved February 13th, 2017, from http://articles.sae.org/11937/.

Behdani, B., Fan, Y., Wiegmans, B., Zuidwijk, R. (2016). Multimodal schedule design for synchromodal freight transport systems, *European Journal of Transport and Infrastructure Research*. 16 (3): 424–444.

Den Hengst-Bruggeling, M. (1999). Interorganizational coordination in container transport. Doctoral thesis, Delft University of Technology.

Fan, Y. (2013). The design of a synchromodal transport system: Applying synchromodality to improve the performance of current intermodal freight transport system. Master's thesis, Delft University of Technology.

Froese, J., Töter, S. (2014). Green and effective operations at terminals and in ports: The outcome. *GreenPort AUTUMN*.: 28–38.

Geerlings, H., van Duin, R. (2011). A new method for assessing CO_2-emissions from container terminals: A promising approach applied in Rotterdam. *Journal of Cleaner Production*. 19 (6–7): 657–666.

Hekkenberg, R., Liu, J. (2016). *Inland Waterway Transport: Challenges and Prospects*, (edited by Wiegmans and Konings). Routledge Studies in Transport Analysis, Routledge. London.

IGU. (2015). World LNG Report: 2015 Edition. Retrieved October 10th, 2016, from http://www.igu.org/sites/default/files/node-page-field_file/IGU-World%20LNG%20Report-2015%20Edition.pdf

Jonkeren, O., Rietveld, P., van Ommeren, J., Linde, A. Te. (2013). *Climate Change and Economic Consequences for Inland Waterway Transport in Europe*. Springer. Berlin.

Konings, R., Kreutzberger, E., Maraš, V. (2013). Major considerations in developing a hub-and-spoke network to improve the cost performance of container barge transport in the hinterland: The case of the port of Rotterdam. *Journal of Transport Geography*. 29: 63–73.

Mommens, K., Macharis, C. (2014). Location analysis for the modal shift of palletized building materials. *Journal of Transport Geography*. 34: 44–53.

NEA. (2009). Kostenkengetallen binnenvaart 2008, Eindrapport, NEA, Zoetermeer.

Plompe, W. (2014). Establising guidelines for port hinterland intermodal inland waterway transport network design. MSc thesis report. Delft University of Technology.

Rotterdam, P. o. (2014, August 12). Rotterdam en Beieren starten intermodaal onderzoek. Retrieved February 10th, 2016, from https://www.portofrotterdam.com/nl/nieuws-en-persberichten/rotterdam-en-beieren-starten-intermodaal-onderzoek.

Smith, L.D., Sweeney, D.C., Campbell, J.F. (2009). Simulation of alternative approaches to relieving congestion at locks in a river transportion system. *Journal of the Operational Research Society*. 60 (4): 519–533.

Veenstra, A., Zuidwijk, R., van Asperen, E. (2012). The extended gate concept for container terminals: Expanding the notion of dry ports. *Maritime Economics & Logistics*. 14: 14–32.

van Duin, J.H.R., Geerlings, H. (2012). Estimating CO2 footprints of container terminal port-operations. *International Journal of Sustainable Development and Planning*. 6 (4): 459–473.

van Duin, J.H.R., van der Heijden, R.E.C.M. (2012). A new barge terminal in a residential area: Using simulation modeling to support governance of noise. *Journal of Computational Science*. 3 (4): 216–227.

van Hassel, E. (2015). *(Over)capaciteitsontwikkeling in de binnenvaarttankermarkt en mogelijke toekomstscenario's*. University of Antwerp, Faculty of Applied Economics. Antwerp.

Wiegmans, B., Konings, J.W. (2015). Intermodal inland waterway transport: Modelling conditions influencing its cost competitiveness. *The Asian Journal of Shipping and Logistics*. 31 (2): 273–294.

Chapter 5

Road distribution from the intermodal perspective

Rickard Bergqvist, Jason Monios and Sönke Behrends

INTRODUCTION

This chapter focuses on the relevance of road haulage from the intermodal perspective. It describes how distribution centres feed intermodal links and also new initiatives such as government policies to allow longer and heavier vehicles to perform the first and last mile. It also includes a discussion of urban distribution, its limitations in terms of intermodal transport, but its increasing importance due to concerns about emissions in urban areas. The overarching aim of this chapter is to highlight the challenges and requirements of integrating pre- and post-haulage (PPH) with intermodal transport for the trunk haul. According to the European Commission (2012), road freight transport accounts for 73% of all inland freight transport in the European Union (EU). Because of the commonly associated negative impacts from road transport, the European Commission suggests that 30% of road freight over 300 km should shift to other modes such as rail or waterborne transport by 2030, and more than 50% by 2050. In order to achieve this goal, the development of intermodal transport is crucial.

Intermodal transport compared with road haulage has some inherent challenges, such as longer transit times, cost, reliability and flexibility and the need to remodel the supply chain to incorporate potential changes such as increased in-transit inventory and extended delivery windows (see Chapter 9). To address and manage these challenges, increased collaboration is needed between key stakeholders, such as shippers, third-party logistics providers (3PLs), hauliers, rail operators and terminal operators. Much supply chain and logistics literature has addressed the key requirements in supply chain and logistics collaboration, such as trust, knowledge sharing, process integration and decision synchronisation, and some authors have found these techniques useful in relation to developing intermodal transport services. Similarly, while significant literature on intermodal transport focuses on the transport cost challenges caused by the rail haul itself and the increased handling costs from changing mode, less research has focused on the last mile (Bergqvist and Monios, 2016; Ye et al., 2014; Caris et al., 2008).

PRE- AND POST-HAULAGE

The predominant modes of transport for the longest links in the intermodal transport chain are rail, inland waterways, short sea shipping or ocean shipping where economies of scale apply. Intermodal transport thus enhances the cost-efficiency of the transport system. Road transport, as part of an intermodal transport chain, is often assigned to short-haul and PPH. Most of the vehicles used in intermodal freight transport by road are either articulated combinations comprising a tractor (i.e. motive unit or traction unit – see Figure 5.1) or a semi-trailer (Figure 5.2).

PPH costs represent one of the most important cost components in relation to the competitiveness of intermodal transport services (e.g. Niérat, 1997; Kreutzberger, 2001). For intermodal transportation over shorter distances (e.g. below 300 km) and where there is a PPH component at both ends of the chain, the competitiveness of the intermodal transport system compared with direct road is weak. It is hard to calculate exact break-even estimates for an intermodal route, but for routes that require no road haulage estimates and cases show that it can be as low as around 90 km. With a road haul at one end only, the break-even increases to roughly 200 km, and if both pre- and end-hauls are required, the distance is approximately 450–500 km (Bergqvist, 2009).

In the intermodal transport chain, PPH operations normally involve the transport of an empty intermodal loading unit (ILU) and the subsequent transportation of a full ILU between the shipper and the intermodal terminal (Macharis and Bontekoning, 2004). Kreutzberger et al. (2006) evaluated the cost performance of PPH and identified some critical factors, including the haulage distance, the freight volumes per shipper or area, the resource productivity, the labour or fuel costs and the network productivity (e.g. number of round trips per load unit and

Figure 5.1 Tractor unit connected to a trailer loaded with an ISO container. (From Rickard Bergqvist, Jason Monios and Sönke Behrends.)

Figure 5.2 Curtain-sided trailer commonly used in road transport. (From Rickard Bergqvist, Jason Monios and Sönke Behrends.)

loading/unloading times). In Europe, most PPH operations related to inland terminals have a distance of 0–25 km (one-way), only a few trips exceed a distance of 100 km, and the number of terminal visits/trips per day of a truck is 1.4–2.1 (Kreutzberger et al., 2006).

PPH operations are usually very fragmented, with many different hauling companies serving each terminal. Delivery and pick-up movements to and from shippers are often not coordinated between hauliers, or even by single hauliers, resulting in many empty trips and a low utilisation rate of equipment (Morlok et al., 1995). On the other hand, the intermodal transport set-up may lead to concentrated PPH flows and can potentially create substantial waiting times at large-scale intermodal terminals (Walker, 1992). Since many intermodal terminals are also located in urban areas, PPH is affected by urban congestion. For rail-based intermodal transport services that are usually very dependent on their lead time, PPH can actually be the primary source of both long transit times and lead-time unreliability (Morlok et al., 1995). PPH activities also account for a large fraction (between 25% and 40%) of the total cost of an intermodal transport service, despite its relatively short distance (Macharis and Bontekoning, 2004).

For PPH activities, there is often a low distance-dependent marginal cost, based on labour time and fuel costs, and a relatively high initial cost, based on the high fixed costs of the truck assets as well as the opportunity cost of allocating a truck to a route. Hence, PPH significantly affects the service quality and competitiveness of intermodal transport and thus limits the markets in which it may compete. Improving the efficiency of PPH can, thus, greatly improve the attractiveness of intermodal transport. Morlok et al. (1995) showed that a decrease of 30% in PPH costs can reduce the break-even distance by as much as 42% for intermodal transport compared with direct road. They also concluded that improvement in

PPH is the key for enlarging the intermodal market, since improvements in other parts of the intermodal transport chain will not lead to the same proportional decrease in costs. However, it is difficult and challenging to achieve improvements to the set-up and organisation of PPH (Niérat, 1997). Changes to the regulatory framework might be one such example of improvement that could be possible and even desirable. Regulations for the transport market determine to a large extent the cost level of PPH. Road transport is connected to a large number of government directives and regulations, including maximum vehicle dimensions and weights, operator licencing, drivers' working times and so on. (Lowe, 2005). Environmental regulations also affect PPH operations since PPH transport often takes place in urban areas where additional regulations may apply.

URBAN DISTRIBUTION AND INTERMODAL TRANSPORT

Intermodal transport has a significant urban dimension. Since intermodal terminals as well as the shippers and receivers of rail freight are often located in urban areas, urban transport constitutes both an important and a limiting factor for modal shift strategies. On the one hand, urban transport increases the reach of rail freight by linking shippers and receivers to rail and enabling the regional consolidation of goods flows. While the volumes of individual shippers are often too small for intermodal transport, the total transport demand of several shippers in the same urban area may be sufficient to achieve volumes which are economically viable for intermodal services. The required volumes therefore can be achieved by consolidating the less-than-trainload shipments from different shippers located in close proximity to each other. A city is usually provisioned by hundreds of supply chains, since it hosts shippers from many economic sectors (Dablanc, 2011). The required regional consolidation is achieved at the intermodal terminals by road transport collecting and distributing the shipments in the terminal region.

On the other hand, the urban environment, characterised by congested roads, space constraints and the limitation of infrastructure, restricts the efficiency of PPH and terminal operations. Freight traffic shares the infrastructure with passenger traffic and hence is affected by urban congestion, which impairs the quality and efficiency of PPH operations. This impedance from urban freight is especially relevant for shorter distances, which are more sensitive to additional transhipment and PPH costs because they represent a greater share of total chain costs (Bontekoning and Priemus, 2004). Hence, urban freight transport is both an enabler and a barrier to the desired modal shift (Figure 5.3).

Furthermore, urban PPH operations, which are vital for the efficiency of intermodal transport, are increasingly perceived as negative activities in respect of passenger transport and the quality of life of citizens. Urban PPH often takes place during rush hour and hence contributes to urban congestion, while single-mode road transport often takes place during the night with no congestion effects. The externalities of PPH compared with single-mode road freight are therefore significantly higher. To compensate for these additional traffic impacts from PPH in

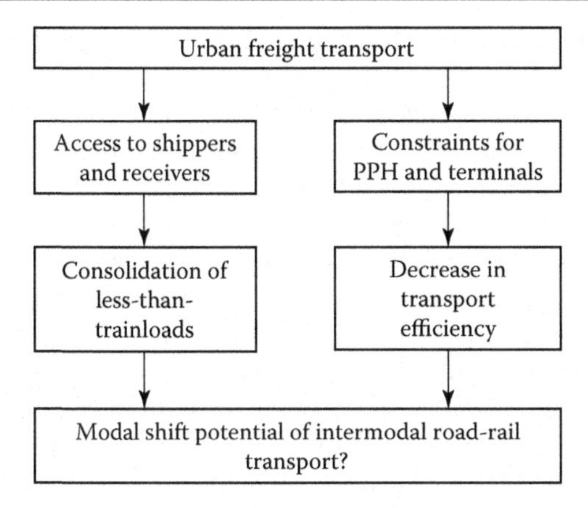

Figure 5.3 The impact of urban freight transport on the modal shift potential of intermodal road-rail transport. (From Rickard Bergqvist, Jason Monios and Sönke Behrends.)

urban areas, a certain distance needs to be covered to achieve enough savings in CO_2 emissions on the long haul. The break-even distance for achieving an environmental benefit depends on the relative advantage of rail over road in terms of climate impact and the relative disadvantage of PPH over single-mode road in terms of traffic impacts. Assuming a further introduction of alternative fuels in the road freight sector, which potentially decreases the environmental benefits of rail on the one hand, and increasing congestion problems in cities, which increases the traffic impacts of PPH on the other hand, the break-even distance for a modal shift is likely to increase in the future. This challenges the possibilities to reduce the CO_2 emissions in the freight sector by modal shift.

Moreover, despite the fact that a modal shift might be beneficial for the wider community in terms of decreasing total emissions in a region or country, it may be a disturbing factor for cities aiming for a high quality of life, as PPH increases the impacts on congestion and air quality in the origin and destination cities. Behrends (2012) showed that in cases of unfavourable geographical conditions of terminal, shipper and receiver in the urban setting, the climate impact and air pollution benefits of rail are achieved in conjunction with higher traffic impacts from PPH operations in urban areas. This can have negative implications for modal shift strategies, since investments in intermodal terminals, which are a prerequisite for the future growth of rail freight, are likely to be opposed by local authorities if a modal shift increases the externalities in urban areas.

The scale of the geographical trade-off is largely determined by the relative location of the intermodal terminal, shipper and receiver in the spatial structure. This structure can vary significantly. While in many European and North American cities distribution and warehousing facilities have been decentralised in recent decades to suburban locations, many of the rail terminals still tend to be located near urban centres. These were originally established in the nineteenth century at the edge of urban areas, but are now surrounded by urban development (Rodrigue et al., 2013). In the current spatial structure of many cities, the intermodal terminals are therefore often located close to the city centre, while the shippers and receivers of intermodal freight are often located at the urban fringe areas with good connections to the surrounding highway network. This urban spatial structure generates PPH flows that may pass through central areas on their way to suburban production and logistics facilities as well as to regional markets, which has negative implications for both the city region's sustainability and the efficiency of the intermodal networks. Behrends (2012) showed that an alternative terminal location closer to the shippers and receivers can significantly decrease the distance of PPH trips in urban areas and hence decrease its traffic impacts. These savings can be substantial and can even result in lower externalities than for all-road, if the terminal and the shipper are located in close proximity to each other. The significantly smaller traffic impacts in urban areas can encourage local authorities, rather than forcing them, to help integrating intermodal transport in the urban spatial structure. As a consequence, Behrends (2012) concluded that short PPH distances between terminal and shipper/receiver within the urban area can reduce the structural disadvantages of intermodal transport over road, with the result that a modal shift for relatively short distance transport can also result in total environmental benefits. Such a rail-adapted land use planning approach, resulting in more environmentally friendly PPH operations, will also be beneficial for the competitiveness of intermodal transport, for which PPH time and costs are crucial factors. Land use planning therefore must consider the interactions of urban freight and intermodal transport instead of handling them as separate policy concerns.

However, despite this significant urban context of intermodal transport, neither urban studies nor intermodal transport research have paid particular attention to this subject. Urban planners and intermodal transport stakeholders usually focus on their own field and neglect the relationships between intermodal transport and urban freight transport. Intermodal transport research is usually limited to rail haulage and transhipments, while PPH in urban areas is regarded as an activity beyond the system boundaries (Woxenius and Bärthel, 2008). In policy and planning, urban freight and intermodal transport are still handled as separate policy concerns, and their interactions are rarely considered (Nemoto et al., 2005). Intermodal transport policymakers and practitioners need to take into account the urban context of intermodal transport when designing strategies for modal shift. The required adaptation to the urban environment does not necessarily result in additional friction but can be considered positive. The urban context also offers opportunities, which can increase the market potential of rail freight, increase the efficiency of PPH and reduce the urban impacts of intermodal transport. Cooperation among actors and integrated land use and transport planning can be significant in this respect. Local authorities can play a key role in enabling cooperation among shippers and transport operators by involving all stakeholders in the strategic land use and transport planning processes.

They, therefore, have an important role to play if a sustainable modal shift is to be achieved.

INBOUND LOGISTICS AND THE LAST MILE

It is important for shippers to be aware that the management of inbound logistics and PPH, in connection with intermodal transport, can achieve great benefit for the whole supply chain. For example, keeping the storage of incoming containers at an inland terminal rather than a seaport means that they are closer to the location of shippers' distribution centres (DCs). This generates benefits by quicker lead times from calling in a container from their forwarder for delivery to their warehouse. However, in order to improve the effectiveness of inbound logistics, close cooperation and integration of the final transport leg between terminal and shippers' DCs is essential. The final leg is the interconnection between the inland terminal and the storage of containers and the final activity of stripping/stuffing the container before the goods are moved into/out of the warehouse. Shippers usually desire a system that enables quick and smooth capacity adjustments and transport planning that makes it possible to change the order of containers, the frequency of delivery and the total throughput on a more or less real-time basis.

From a transport system perspective, one of the biggest challenges is to optimise the opening hours of the inland terminal with the much longer opening hours of many shippers' DCs. The inland terminal may be flexible with opening hours, but at an extra cost. The other option is to acquire more road trailers and let the terminal operator load the extra trailers before closing, so that the haulier can continue PPH operations outside the terminal's opening hours.

To optimise the total cost of the final part of the intermodal transport chain, for example storage of containers, the final PPH leg, stripping/stuffing of containers and empty repositioning of containers, the shipper, haulier and the terminal operator need to identify a number of parameters that should be adjusted to achieve minimum total cost for the inbound logistics. The main parameters (Bergqvist and Monios, 2016) are:

- Number of containers stored at the terminal
- Opening hours of the terminal
- Number of traction units (i.e. trucks) and working hours of drivers
- Capacity of trucks (e.g. high capacity transport [HCT] – see section on high capacity transport)
- Number of trailers
- Opening hours and manning of the central warehouse

Since the system involves a lot of rapid and unpredicted changes, warehouse capacity, inventory levels and so on, it is necessary to develop a system that is both cost-efficient while at the same time providing high service levels by means of high flexibility. Another complicating factor is that inbound flows are very

Table 5.1 Example system design characteristics

	Intermodal terminal	*Road haulage*	*Distribution centre*
Capacity	• 1 reach stacker	• 2 trucks • 22 trailers (either 1*40 ft or 2*20 ft per trailer) • 5 lorry drivers	• Between 12 and 24 gates • Staffing 2 shifts of 14 on average per shift
Opening/ operating hours	Mon–Fri 05.00–18.00	Mon–Fri 04.30–23.30	Mon–Thu 06.00–23.00 Fri 06.00–16.00 Sat 06.00–20.00
Activity and lead times	• Loading/unloading time: 2–5 min per trailer • Outside opening hours: • Time for marshalling of trailer including leaving/picking up trailer at the gate: 20–35 min	Transport time: 30–35 min (one-way).	Time for marshalling of trailer including leaving/picking up trailer at the gate: 20–35 min

seasonal and are highly dependent on peaks in demand at certain times such as Christmas.

Adjustments that could be made to the system, usually through close cooperation between partners, include keeping the parking area and the gates accessible to the road haulier at any time, even if stripping of containers is only carried out, for example, during 06:00–23:00 Monday–Thursday, 06:00–16:00 Friday and 06:00–20:00 Saturday. Table 5.1 summarises the main characteristics of a sample system design.

In Europe, a typical intermodal haulage pattern for inbound flows is for a truck to arrive at the terminal with a trailer, pick up the full container from an inbound train and deliver it to the customer. The truck could back up to a loading bay where the container is emptied while the driver waits (just as in regular non-intermodal truck transport), and the truck then returns the empty container to the empty depot at the terminal. In other cases, the container will be left with the customer. This can only happen if the receiver has their own loading equipment. Moreover, unless they have a ground-level warehouse, the container will need to be transferred onto a raised platform or a dolly to meet the loading bay height. It is clear that there is a built-in delay to this system. This 'grounded' model is common in Europe, whereas in the United States, 'wheeled' operation is common, whereby the container stays on the trailer (referred to as a chassis in the United States). The driver arrives at the terminal with only a tractor unit, hooks up to a waiting chassis with the full container on it and delivers that to the customer, where it is unhooked and the driver leaves with only the tractor unit. This provides more flexibility as the driver does not have to wait, and additional chassis can be lined up at additional loading bays (subject to sufficient loading bays). Adopting this model in Europe can be described in a number of steps:

1. The truck arrives at the terminal with a trailer with an empty container from the previous trip, or with no container. Before arriving, the driver has used a two-way radio to contact the driver of the reach stacker to let him or her know that the truck is on the way, and if necessary to communicate the order of containers that will be picked up. By doing so both at the terminal and the DC, the truck driver can ensure that there is no or limited waiting time.
2. Truck marshalling at the terminal. Truck disconnects the trailer with the empty container(s) and connects to a trailer with full containers.
3. Transport between inland terminal and DC.
4. Truck arrives at the DC. Before arriving, the driver has contacted personnel at the shipper with com-radio to get information regarding which gate to dock the trailer. By doing so, personnel at the DC can try to allocate the goods to the gate in the warehouse closest to the designated storage area proposed by the warehouse management system (WMS). Personnel from the DC open the doors of the container before docking it to the gate in order to save time for the driver.
5. Truck connects to trailer with empty containers.

Two of the most decisive factors related to the performance of the system are the lead times and number of trips required per day given the demand. If the shipper has an open-book agreement with the road hailer, they have a joint ambition to decrease the number of hours spent on haulage combined with the amount of equipment (trucks, trailers, etc.) used. By adding more trailers to the system, it is possible to decrease unloading and loading times at the inland terminal, since the truck drivers do not have to wait for the reach stacker to handle their containers and trailers, as new trailers were already waiting and could just be picked up.

Furthermore, more trailers means that the haulage company can work with a smaller time window than the personnel stripping containers at the DC, since the truck drivers are able to accumulate full containers on spare trailers at the DC during their working hours (throughput of the haulage per time-unit is higher than that of the stripping personnel). Also, the total throughput of containers can be increased as the inland terminal operator can prepare more trailers that can be handled outside the regular opening hours, since the trucks have access to the area where trailers are handled at the inland terminal. This permission is granted by the terminal operator, as they recognise the huge importance of this large shipper to the viability of the terminal. When increasing the number of trailers substantially, it is even possible for the personnel at the DC to strip containers by just handling the full containers accumulated during the week, without requiring either the haulier or the inland terminal operator to work during the weekends, which would incur higher hourly rates.

With regard to future developments, one aspect that is currently discussed is the trade-off between terminal opening hours and the extra time for marshalling required after opening hours by the road haulier. For every round trip, the road haulier could save 15–20 min if marshalling were avoided. For a normal day, extending the opening hours of the terminal could potentially save 20 min each for eight trailers (four round trips), which would mean a daily saving of 80 min for the road haulier. However, this saving needs to be balanced against the extra cost for the increased opening hours of the terminal.

HIGH CAPACITY TRANSPORT

There are typically three types of ILU: swap bodies, semi-trailers and International Organization for Standardization (ISO) containers (see Chapter 2). The most common dimensions of ISO containers are 20, 40 or 45 ft long (5.98, 12, 13.50 m). For swap bodies, two classes can be distinguished. For carriage on road trains, 'Class C' swap bodies with lengths of 7.15, 7.45 and 7.82 m are used. For articulated vehicles, 'Class A' swap bodies with lengths of 12.50 and 13.60 m are the most common. For semi-trailers, the typical length is 13.60 m (Vrenken et al., 2005).

One of the most notable improvement projects related to transport in Europe is HCT, commonly used as a generic term for any vehicle combinations and dimensions that exceed existing regulations on either length and/or weight. Specific formats such as longer heavier vehicles (LHV – sometimes also known as mega truck or eco combi) have been explored in recent years. The European Union defines LHV as 'all freight vehicles exceeding the limits on weight and dimensions established in Directive 96/53/EC'. The maximum dimensions of vehicles as well as the weight limits in national and international road freight traffic are regulated by individual countries as well as cross-border agreements such as those in place in the European Union. EU Council Directive 1996/53/EC restricts vehicle lengths on cross-border traffic to 16.50 m for truck-trailer combinations and to 18.75 m for articulated vehicles. The maximum permissible weight is 40 t and can only be extended to 44 t when transporting containers to and from intermodal terminals. Individual countries may set their own limits. Some countries (e.g. Sweden, Finland and the Netherlands) allow 25.25 m length vehicles, although with differing weight restrictions. While each of these regulations is set within individual countries, the European Union is considering a new regulation to allow cross-border transport of LHVs of 25.25 m length and 60 t weight (Sanchez Rodrigues et al., 2015).

LHV combinations are possible through the European modular system (EMS). This system is used in Sweden and Finland and under trial in Belgium, Denmark, Germany and the Netherlands (OECD, 2010). Council Directive 1996/53/EC allows member states to legalise longer and heavier vehicles, so long as they conform to the standard modular dimensions defined in the directive. The short module 7.82 m, which is a European Committee for Standardisation (CEN)

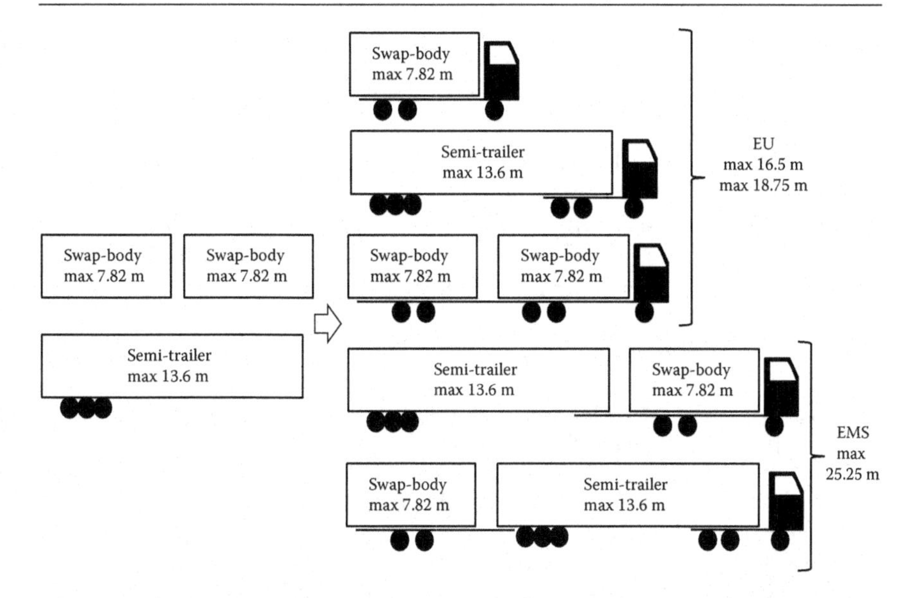

Figure 5.4 Variety of road trains with different semi-trailer/swap-body combinations (EMS). (From Rickard Bergqvist, Jason Monios and Sönke Behrends.)

standard for swap bodies, also includes other standardised load units, such as 7.45 m, 7.15 m and 20 ft. The long module 13.6 m, which is the European semi-trailer length, includes the 40 ft ISO container. These vehicle units can be coupled together to achieve a total loading length that is a multiple of the module lengths 7.82 m and 13.6 m. Figure 5.4 illustrates different combinations of the EMS.

Various studies have been undertaken with the aim of analysing the relative economic and environmental costs and benefits of longer road vehicle combinations. In Sweden and Finland, experience of using EMS vehicle combinations is mostly positive (Åkerman and Jonsson, 2007). The study by Åkerman and Jonsson (2007) concluded that the use of LHVs with EMS in Sweden and Finland has had a positive effect on the economy and environment without negatively affecting traffic safety. Potential negative effects are the generation of new freight traffic if companies respond to the reduction in freight costs by increasing their freight flow, a possible increase in the severity of accidents as a result of the greater weight and size and a possible increase in infrastructure costs of road to accommodate LHVs (McKinnon 2008). Hence, increasing the maximum length and weight of road vehicles is a controversial issue in the context of transport policy. McKinnon (2008) also highlights the difficulties in assessing the net benefits and extrapolating the experience from single national trials with unique national characteristics to the European Union as a whole.

However, recent research shows that there might be interesting potential in allowing for longer but not heavier vehicle combinations to improve the

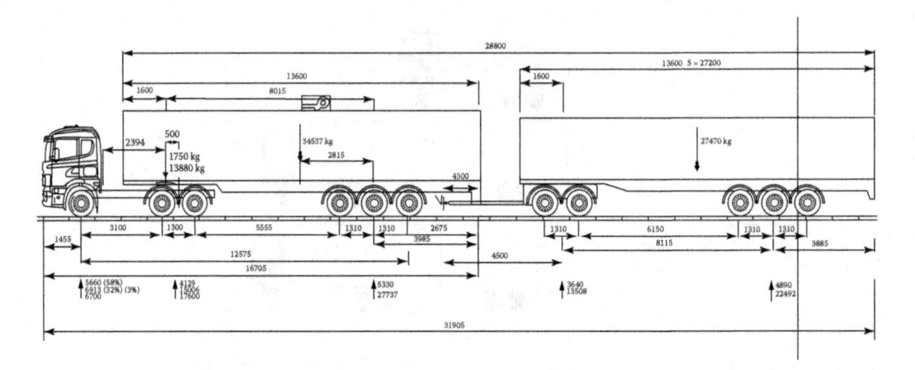

Figure 5.5 Example of vehicle for haulage of 2*40 ft containers. (From Rickard Bergqvist, Jason Monios and Sönke Behrends.)

competitiveness of intermodal transport. As previously described, in many countries, regulations allow the transportation of either one 40 ft or a 40 ft + 20 ft combination. Bergqvist and Behrends (2011) found that by altering this model to 40 ft + 40 ft for PPH (see Figure 5.5) to and from intermodal terminals, savings of 5%–10% are possible. One of the reasons for the large saving is the current imbalance that many shippers have between 20 ft and 40 ft containers, meaning that they cannot fully utilise regulations on vehicle combinations since they usually have more 40 ft than 20 ft containers (Bergqvist and Monios, 2016; Ye et al., 2014; Bergqvist and Behrends, 2011). It is important to repeat that, in this case, the regulatory change to vehicle restrictions has been related to length of vehicles and not the maximum payload of vehicles, which has a greater impact on road safety and infrastructure restrictions and limitations.

CONCLUSION

This chapter has considered road haulage from the intermodal perspective, including a perspective on the necessary urban characteristic of this element of the intermodal transport chain. The chapter provided an overview of road haulage practices to and from intermodal terminals, the role of regulation and new initiatives to allow longer and heavier vehicles to perform the first and last mile. The key contribution of this chapter was to describe the challenges and requirements of integrating PPH with intermodal transport for the trunk haul.

REFERENCES

Åkerman, I., Jonsson, R. (2007). European modular system for road freight transport: Experiences and possibilities. Report 2007:2 E, TransportForsk (TFK), Stockholm, Sweden.

Behrends, S. (2012). The significance of the urban context for the sustainability performance of intermodal road-rail transport. *Procedia: Social and Behavioral Sciences*. 54: 375–386.

Bergqvist, R. (2009). *Hamnpendlars Betydelse för Det Skandinaviska Logistiksystemet*. BAS Publishing, Göteborg, Sweden.

Bergqvist, R., Behrends, S. (2011). Assessing the effects of longer vehicles: The Ccase of pre- and post-haulage in intermodal transport chains. *Transport Reviews*. 31 (5): 591–602.

Bergqvist, R., Monios, J. (2016). Inbound logistics, the last mile and intermodal high capacity transport. *World Review of Intermodal Transport Research*. 6 (1): 74–92.

Bontekoning, Y.M., Priemus, H. (2004). Breakthrough innovations in intermodal freight transport. *Transportation Plannning and Technology*. 27 (5): 335–345.

Caris, A., Macharis, C., Janssens, G. K. (2008). Planning problems in intermodal freight transport: Accomplishments and prospects. *Transportation Planning and Technology*. 31 (3): 277–302.

Dablanc, L. (2011). City distribution, a key problem for the urban economy: Guidelines for practitioners. In Macharis, C., Melo, S. (Eds.), *City Distribution and Urban Freight Transport*, pp. 13–36. Edward Elgar, Cheltenham, UK.

European Commission. (2012). Review of EU rules concerning access to the EU road haulage market and access to the occupation of road transport operator. DG MOVE Unit D.3. Retrieved March 24th, 2017, from http://ec.europa.eu/smart-regulation/impact/planned_ia/docs/2013_move_003_eu_road_haulage_market_en.pdf.

Kreutzberger, E. (2001). Strategies to achieve a quality leap in intermodal rail or barge transportation. IEEE Intelligent Transportation Systems Conference Proceedings, 25–29 August 2001, pp. 785–791. Oakland, CA.

Kreutzberger, E., Konings, R., Aronson, L. (2006). Evaluation of pre- and post haulage in intermodal freight networks. In Jourguin, B. Rietveld, P., Westin, K. (Eds.), *Towards Better Performing Transport Networks*, pp. 256–284. Routledge, London.

Lowe, D. (2005). *Intermodal Freight Transport*. Elsevier Butterworth-Heinemann, Amsterdam.

Macharis, C., Bontekoning, Y. M. (2004). Opportunities for OR in intermodal freight transport research: A review. *European Journal of Operational Research*. 153 (2): 400–416.

McKinnon, A. (2008). Should the maximum length and weight of trucks be increased? A review of European research. Paper presented at 13th International Symposium on Logistics (ISL 2008), 6–8 July 2008, Bangkok, Thailand.

Morlok, E. K., Sammon, J. P., Spasovic, L. N., Nozick, L. N. (1995). Improving productivity in intermodal rail-truck transportation. *The Service Productivity and Quality Challenge*. 5: 407–434.

Niérat, P. (1997). Market area of rail-truck terminals: Pertinence of the spatial theory. *Transportation Research Part A: Policy and Practice*. 31 (2): 109–127.

OECD. (2010). *Moving Freight with Better Trucks: Improving Safety, Productivity and Sustainability*. OECD Publishing, Paris.

Rodrigue, J.-P., Comtois, C., Slack, B. (2013). *The Geography of Transport Systems*. Hofstra University, Department of Global Studies and Geography, Hepmstead, NY.

Sanchez Rodrigues, V., Piecyk, M., Mason, R., Boenders, T. (2015). The longer and heavier vehicle debate: A review of empirical evidence from Germany. *Transportation Research Part D*. 40: 114–131.

Vrenken, H., Macharis, C., Wolters, P. (2005). *Intermodal Transport in Europe*., European Intermodal Association, Brussels.

Walker, W., T (1992). Network Economies of Scale in Short haul truckload Operations. *Journal of Transport Economics and Policy*, 26, 3–7.

Woxenius, J., Bärthel, F. (2008). Intermodal road-rail transport in the European Union. In Priemus, H., Nijkamp, P., Konings, R. (Eds.), *The Future of Intermodal Transport: Operations, Design and Policy*, pp. 13–33. Edward Elgar, Cheltenham, UK.

Ye, Y., Shen, J., Bergqvist, R. (2014). High capacity transport associated with pre- and post- haulage in intermodal road-rail transport. *Journal of Transportation Technologies*. 4 (3): 289–301.

Chapter 6

Intermodal terminal design and operations

Rickard Bergqvist

INTRODUCTION

Intermodal terminals are a crucial part of the transport system and the development of sustainable transport, since they enable the combination of more environmentally friendly and cost-efficient modes of transport. This chapter describes the process of developing intermodal terminals (primarily road-rail intermodal terminals), including the layout of an intermodal terminal, and covers all the decisions about location, track length, electrification, transfer yard, storage space, marshalling and so on. It then continues to analyse the operations of handling, stacking, dealing with train operators, dealing with trucks arrivals and so on. A brief discussion of the terminal's role in the network is included, and a framework for situating analysis within the life cycle of intermodal terminals is also incorporated.

INTERMODAL TERMINAL DEVELOPMENT

Developing intermodal terminals is a complex process, including many actors and functions. The development process can look quite different depending on the actors involved, who takes overall responsibility and so on. (Bergqvist et al., 2010). The most common developers are the different levels of government (local, regional or national), real estate developers, transport operators, third-party logistics providers (3PLs), port authorities, port terminal operators, shipping lines and independent terminal operators (Monios, 2014). The type of actors involved (e.g. public/private) greatly affects the initiative and logic behind terminal development. Public actors tend to focus on social and economic benefits, such as employment and attracting new establishments, while private actors such as real estate developers may see the terminal development as an investment opportunity in itself and have the ambition to sell the site or part of it for profit. Port authorities and port terminal operators tend to see inland terminal development as a strategic means for better hinterland capture.

There is one major strategic aspect separating terminals into two very different types of categories: an independent terminal servicing the traffic of different stakeholders or terminals that operate on the basis of exclusivity for their own needs and purposes. This chapter focuses primarily on the category of independent open-access terminal development and design. This type of terminal often has a greater challenge, initially, in gaining a large enough critical volume for profitable operation (Bergqvist et al., 2010). Private sector terminal developments tend to focus more on a logistics platform since real estate actors are the primary developer. This is common in countries where the public sector has less direct involvement, such as the United Kingdom and the United States (e.g. BNSF Logistics Park Terminal, Joliet, Illinois). Europe, traditionally, has had publicly managed rail networks until recently (Martí-Henneberg, 2013), which means that terminals are developed by both private transport operators and national public rail operators. Figure 6.1 illustrates an intermodal terminal (centre) and a logistics platform (to the left of the terminal).

According to previous research and studies on terminal development (e.g. Bergqvist et al., 2010; Höltgen, 1995; Roso, 2009; Bergqvist, 2007), there are some key success factors when developing intermodal terminals:

- Market potential
- Location
- Entrepreneurship
- Large shipper
- Financiers

Market potential is the most important factor since it will determine the economic viability of the terminal; this is true for both private and public developments. In the case of a private developer, it is often a rail operator or similar industry actor that initiates the terminal development, since they want to develop a specific intermodal transport solution based on their particular demands and needs. For public sector developers, a local market potential might be identified,

Figure 6.1 Illustration of intermodal terminal and logistics platform. (From Jernhusen AB.)

but no direct business commitments can be guaranteed in advance, which means that the public sector might take a greater initial risk before possibly handing the operations to the private sector eventually.

The location of intermodal terminals is a paramount decision since it affects the effectiveness and efficiency of the terminal and ultimately also the competitiveness of the intermodal services (Bergqvist and Tornberg, 2008). This aspect is too often neglected since many, especially first-time developers, do not have a clear understanding of the effect that the terminal location might have on overall rail transport productivity; it is not just about the geographical location but also how and where the terminal connects to the rail network. Therefore, the cost of development usually grows significantly since a number of investments need to be made to ensure that the productivity of the terminal is sufficient and satisfactory. Examples of aspects that affect location include whether or not the nearby rail line is single or double track, in which direction most of the traffic to and from the terminal will pass and the location of existing switches. Another important factor related to location and the cost of terminal development is the topography of the site, since moving and removing large amounts of earth are very costly. Because it incorporates rail infrastructure, a terminal needs to have a very gentle slope. Also related to the aspect of location are public plans, which can take years to get approved, as they might be appealed through the public planning system. When there are several locations to choose from, some might be in other administrative regions such as municipalities; there might be competition between municipalities since terminal development is associated with economic development (Bergqvist, 2008). This can happen in both private and public development; for private developments, there might be public subsidies and grants involved leading to competition among public development agencies.

Entrepreneurship requires creativity and persistence in developing an intermodal terminal since the development process usually takes 5–10 years from idea to operation. The entrepreneur is the key stakeholder that drives and pushes development, something that might be important if the process has come to a standstill for reasons such as funding, political disagreements and so on. This is especially important in the early stages of development when there might be just a few actors involved and no formal agreements signed.

From the perspective of entrepreneurs, the development process needs individuals that can help to establish credibility among the main stakeholders, such as transport authorities, shippers, transport service providers, terminal operators and 3PLs. Large shippers can play an especially important role in this regard, since they can also provide a substantial local market and the demand that contributes to the possibility of intermodal transport services using the terminal. Letters of intent and memoranda of understanding represent ways to formally build credibility.

Funding and finding financiers is another important factor that affects many aspects of the development process and the collaborative atmosphere. Private and public financiers can have very different time horizons for their investment and commitment, which can be difficult to combine since they might have different views on commercial conditions and how quickly they expect a return on their investment. From a development process perspective, it is important to involve financiers with previous experience and knowledge of transport infrastructure investment (Bergqvist, 2009). Due to the long-term development process, the number of stakeholders involved and the different strategic perspectives, coordination throughout the development process is challenging. Challenges in coordination also include issues such as unequal distribution of costs and benefits, lack of resources, different strategic considerations, lack of a dominant actor and risk-averse behaviour (Van der Horst and de Langen, 2008). However, several solutions can be applied (Bergqvist and Pruth, 2006; Bergqvist, 2012):

- *Incentives.* By introducing collaborative incentives, for example bonuses, penalties, tariff differentiation, warranties, capacity regulations, deposit arrangements and tariffs linked to cost drivers, the collaborative structure can be highlighted and formalised.
- *Creating collective action.* Introducing public governance facilitates long-term focus and stability in a context that normally might be uncertain and unstable.
- *Formalisation.* Formalisation limits risk on how uncertainties will be addressed. By formalising cooperation, communication, trust and commitment are facilitated. Examples include sub-contracting, project-specific contracts, defined standards for quality and service, formalised procedures and, especially, tendering and concession agreements.

The previously mentioned solutions take place within the scope of an agreed business model established by the initiators of the terminal development and involve a combination of public and private investments and responsibilities.

INTERMODAL TERMINAL DESIGN

A number of important general issues need to be considered when deciding on a terminal site and designing an intermodal terminal, in order to ensure efficient rail production, terminal operations and interaction between the terminal and the associated logistics platform (Bergqvist, 2012):

- Location in relation to superior infrastructure. As previously discussed, the location and conditions of the terminal site in relation to the rail infrastructure greatly affect the way that the terminal can be connected to the rail network. Examples of key conditions are location of switches, direction of traffic, closeness to stations and other rail depots and topography.
- Marshalling. It is essential for a terminal to have marshalling facilities in connection to the terminal. This provides efficient switching/marshalling to and from the terminal and prevents unnecessary rail movements.
- Slopes in the terminal area and connecting tracks. This aspect affects capacity,

productivity and investment needs, since large slopes might require different equipment and safety measures (the slope of the connecting rail line and the terminal should not be more than a few per mille [‰] preferably).

- Management of waste water in the area. An intermodal terminal usually measures at least 40,000 m^2, which means that it will accumulate a lot of waste water from rain and melting snow and, therefore, requires a well-dimensioned system for taking care of and removing waste water from the terminal.
- Electrification of the tracks and terminal. If the main rail network is electrified, the terminal itself and the connecting track should be electrified to avoid the need for a diesel shunting locomotive. This saves time, costs and is more environmentally friendly.
- Signalling systems connected to the terminal, the need for switches and so on. The terminal and its connecting track and signals should be, if possible, connected to the main traffic system managed by the public transport authorities since this will ensure better traffic planning and remote 24/7 service with minimal need for manual handling of signals and switches.

Besides the previously mentioned considerations related to the general efficiency and productivity of the rail operations, there are some important issues related to the terminal design itself (see Figure 6.2 for an illustration of a terminal layout):

- Paving. The material used for the terminal surface greatly affects the cost of terminal operations and the maintenance cost of the terminal. The most common material is asphalt; however, asphalt has more expensive maintenance costs and a shorter technical lifetime than concrete stones. Asphalt also increases the wear and tear of truck tyres significantly, which makes terminal operations costlier.
- Truck entry and exits. To facilitate terminal capacity, it is important to consider the flow of vehicles in the terminal, so that there are limited disturbances between trucks and terminal handling equipment. The terminal should also provide enough space for vehicle circulation to avoid unnecessary movements that increase lead times and can potentially generate vehicle congestion.
- Lighting. The lighting at a terminal is a central part of the infrastructure that should be coordinated with other functions such as security modules and outlets for refrigerated containers/trailers.
- Local road network. If all streets within the logistics platform and the terminal can be classified as 'internal' streets, vehicles do not need to follow normal road regulations. This enables heavier, longer vehicle combinations and more cost-efficient vehicles such as tugmasters to perform pre- and post-haulage (PPH).
- Security. Security is a major issue at terminals since terminals usually hold

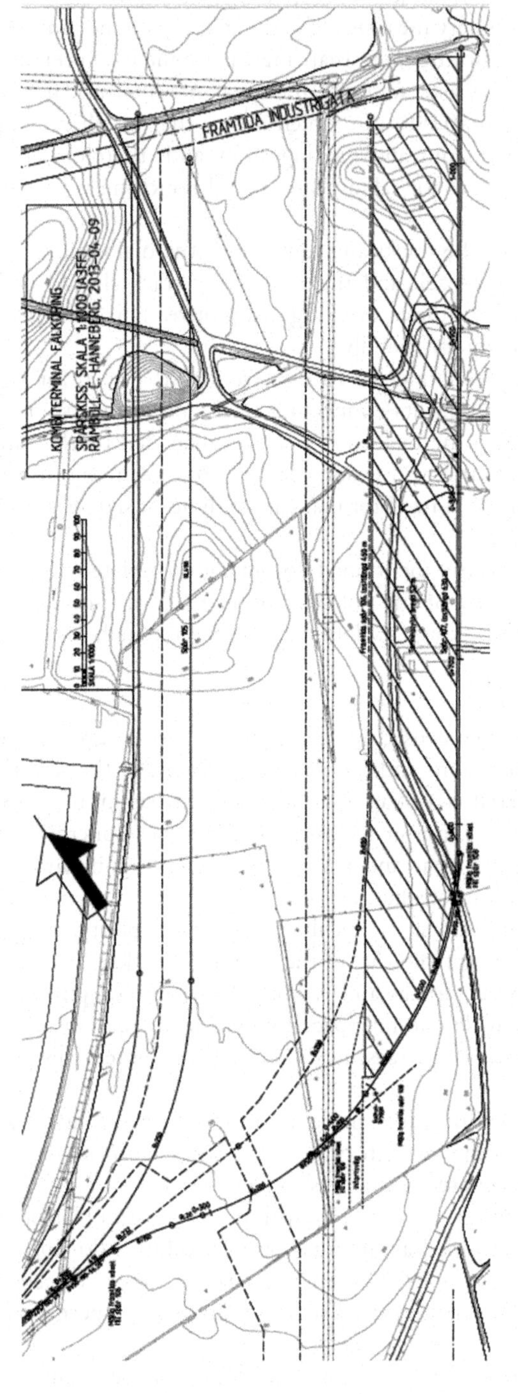

Figure 6.2 Example of blueprint of an intermodal terminal. (From Municipality of Falköping, Sweden.)

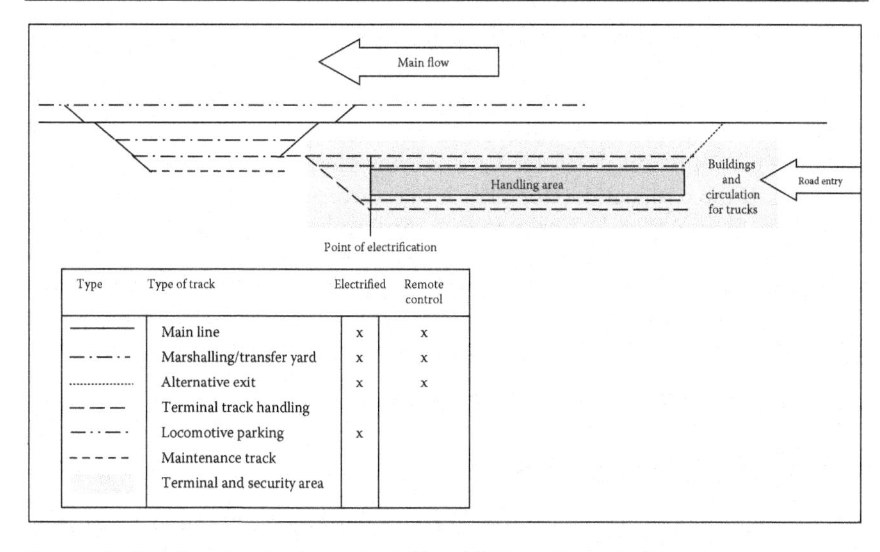

Figure 6.3 The functional unit terminal. (From Rickard Bergqvist.)

significant value due to the goods stored in containers and trailers. A well-functioning security perimeter around a terminal improves security and prevents damage, theft and vandalism, which helps keep insurance down. If a terminal can be planned taking into consideration natural obstacles such as creeks, neighbouring fences and buildings, it can contribute to improving security while also reducing the need for investment in the security perimeter.

Regarding terminal layout, there are some important functions to consider. Figure 6.3 illustrates a standard layout of a functional intermodal terminal.

In contrast to the common European terminal layout, where containers that are not transferred directly from train to truck are stacked on the ground, the common model in the United States is referred to as *wheeled* rather than *stacked*. In the United States, both containers and trailers (referred to as *chassis* in the United States) are owned by the carrier (usually, but not always, the shipping line); therefore, the truck driver simply arrives with a tractor unit. Containers are loaded onto waiting chassis, and the arriving driver hooks up to a loaded chassis and takes it away. These wheeled facilities require a great deal more land space as less equipment can be stacked, but they can be quicker for the incoming drivers who do not have to wait for their container to be located in a stack. This also means that cranes make fewer unproductive moves to pick through a stack of containers.

Again, the importance of the transfer yard for rail transport productivity cannot be stressed enough. The transfer yard should have the following characteristics:

- A yard where groups of wagons can be collected and returned through the effort of only the locomotive crew.
- Remote controlled from the central train control, if possible.

- Electrified, if the connecting line is electrified.
- Sufficient length given the maximum length of trains in the rail network.

The transfer yard and its associated connection to the rail network is an expensive part of terminal development (usually more expensive than the terminal itself). If the terminal development is open access, public subsidies are often available to fund part of the investment. A small-scale intermodal terminal with a capacity of about 30–50.000 twenty-foot equivalent units (TEU) annually would cost around €3–5 million for the terminal itself, and the connecting infrastructure and transfer yard, if they do not already exist, would cost about €3–10 million (depending on factors such as switches, signalling, electrification, topology, planning permissions, cost of land, etc.).

TERMINAL CONCESSION

When evaluating the design and operational model of an intermodal terminal, the most important aspects to consider are whether or not the system should be opened or closed and who should operate the terminal. It might be problematic if too much of a terminal operator's own interests are connected to the associated logistics flows and users of the terminal, as this may affect the market's view of the independence of the terminal operator and the overall credibility of the terminal and its associated services (Bergqvist, 2012, 2013). Although there might be clear conditions in the terminal concession contract, commercial and informational barriers may limit competition. It might be tempting for public actors to operate the terminal themselves if they have invested in the terminal and to avoid the tendering of terminal operations; however, the same actor owning and operating a terminal can distort competition if a public actor does not work under normal profitability requirements. An additional challenge is the transparency required for the role of infrastructure owner and that of terminal operator and the associated performance and quality of service. Even if ownership by public actors is only a part-ownership of the terminal operating company, it can still create issues with transparency and credibility. A clear distinction, where the public actor focuses on ownership and the private sector focuses on terminal operations, is thus preferable, normally categorised as the landlord model and used extensively in the port sector.

Public tendering of terminal operations is preferable to facilitate open and transparent competition and generate market credibility for the terminal and its operations. The next section will elaborate on the issue of tendering of terminal operations. An important advantage of a clear tendering and concession contract is that the infrastructure owner can formalise the monitoring system and conditions and deviations that can lead to the cancellation of a contract, beforehand. These conditions might otherwise be very difficult to implement if the terminal operator has 'possessory rights' to the terminal in a lease agreement or similar.

The tendering process with its associated documentation needs to be well structured and contain detailed information related to services and service levels, contract options, leases, marketing, contract periods, risk sharing and so on. At the same time, the tendering process should be open to new ideas and concepts presented by bidders; this is important to bring in private sector expertise. In this regard, the tendering process can facilitate innovation and creativity in terminal operations and intermodal transport services. During the last decade, much attention and interest has focused on issues related to concessions of ports. Although much research has been done on port concessions, which nowadays are well understood, intermodal terminal concessions and contracts have received less attention. There is little standardisation to be found in relation to contracts and tendering of intermodal terminal operations, procedures, requirements, risks, incentives and so on (Bergqvist and Monios, 2014). Monios and Bergqvist (2015) established a new framework for such analysis based on the Port Reform Toolkit published by the World Bank in 2001 and updated in 2007. The Port Reform Toolkit aims to support the public sector in 'choosing among options for private sector participation and analysing their implications for redefining interdependent operational, regulatory, and legal relationships between public and private parties' (World Bank, 2007: p. xviii). The World Bank Toolkit contains a list of 11 sections normally found in a concession agreement (Table 6.1).

The sections in the Port Reform Toolkit cover basic conditions, such as handover and hand back procedures when changing operator, financial and legal issues, operations, fee-structures and levels and performance monitoring. Monios and Bergqvist's (2015) study of Swedish terminal concession agreements found them to be far less substantial compared with what is suggested

Table 6.1 World Bank toolkit

No.	*Section*
1	Introduction and basic conditions
2	Handover
3	Project control and finance
4	Extension works
5	Operations
6	Fees
7	Legal and insurance
8	Hand back
9	Legal and insurance
10	Performance
11	Legal and insurance

Source: Adapted from Monios, J., Bergqvist, R., *Research in Transportation Business & Management* (*RTBM*), 14 (March). 1–3, 2015.

in the World Bank port concession framework. The weakest sections, often missing or lacking detail, were the sections on extension works, hand back procedures and performance monitoring. This potentially poses major problems since the lack of details and specifications may lead to operational disturbances and uncertainties (Bergqvist and Monios, 2014). To overcome such difficulties and shortcomings, continuous communication is necessary between the public infrastructure owner and the private operator. Some of the more severe information gaps in many of the agreements were the lack of open-access definitions and information on how the terminal and its services should be developed, marketed and promoted. Other examples of lack of information included which performance goals and key performance indicators (KPI) would be used and on what grounds the contract could be terminated, related compensation and so on. Many of these gaps can be explained by the lack of experience of public sector officials managing the process.

TERMINAL OPERATIONS AND GOVERNANCE

The operational phase of an intermodal terminal hopefully lasts for decades over which time the investment is repaid. Over time, variable costs tend to increase as a result of increased demand for maintenance, other upgrading of equipment and so on. The major concern of operations is to continue to attract users to the terminal. This is best done over time by ensuring cost-efficient and high-quality services. The economics of an intermodal terminal is a key focus area in research related to intermodal transport and terminals (see Chapter 8). The costs of an intermodal terminal may be relatively low in comparison to other costs, such as locomotives, wagons and so on, but they are of great importance since it affects the competitiveness of the intermodal service, and it can be improved by means of effectiveness and efficiency. Cost structures related to rail-based intermodal transport have been extensively analysed in other literature (e.g. Janic, 2007).

To ensure overall effectiveness and efficiency, well-functioning terminal governance is needed. One common development model is where the public sector is involved early in the development process, and then private operators eventually operate the terminal, either taking over full ownership or operating the terminal on a concession contract (Monios, 2015). Research shows, however, that terminals developed under such circumstances might have a higher risk of optimism bias (Bergqvist et al., 2010). From the start, it is important to understand the connection between the logistics model/logic of the intermodal terminal and the business model of the terminal (see Chapter 8). Connecting the financial support for developing intermodal terminals with the conditions for the operational model of the terminal is challenging. Therefore, there is always a risk that the terminal may not operate on a viable economic model. This may lead to terminal operation that is difficult to sustain economically or to a situation

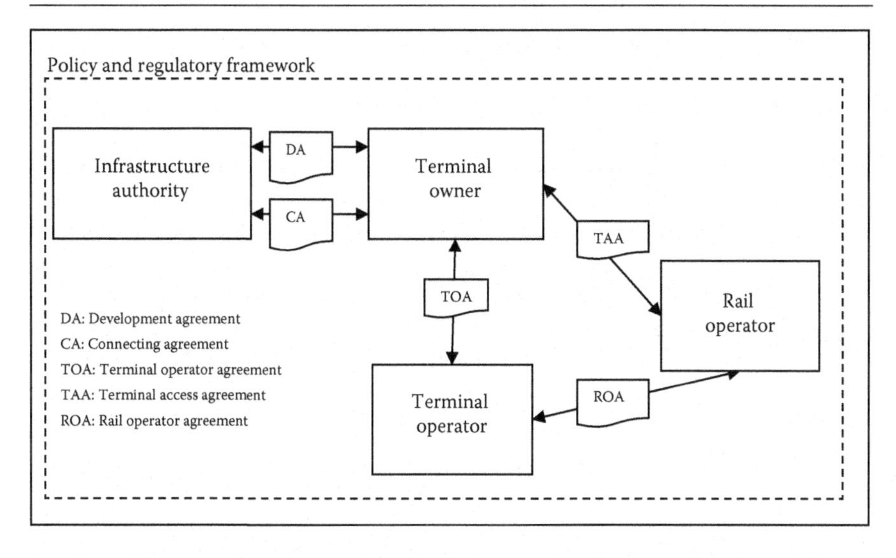

Figure 6.4 Conceptual framework of intermodal terminal governance and contracts. (From Bergqvist, R., Monios, J., *World Review of Intermodal Transportation Research*, 5 (1), 18–38, 2014.)

where the terminal owner might be forced to continuously subsidise the terminal. The business model of the terminal operator is therefore as important as the intermodal terminal services that use the terminal. Given the many stakeholders involved, clear interface between all stakeholders is essential. Figure 6.4 illustrates the different stakeholders involved and the corresponding agreements that are usually in place.

The following is an explanation of the agreements presented in Figure 6.4:

- *Development Agreement* (DA). This agreement stipulates the development of the intermodal terminal and the premises for the development. It defines issues such as who is responsible for analysing what; the different phases of the development; and, most importantly, what is funded by the different actors and are there any conditions connected to the funding that later affect other agreements in the context of the intermodal terminal.
- *Connecting Agreement* (CA). Here, the main component is the definition of the boundaries and responsibilities of the two parties in relation to the developed infrastructure. The agreement is often very technical in nature, with a focus on responsibilities in relation to maintenance. It may also contain the condition that the terminal owner also has the responsibility to modernise the terminal if new technology is introduced by the infrastructure authority. Sometimes, the agreement also contains clarification of who is responsible for traffic planning and management. If the infrastructure authority is responsible, there is usually a cost defined for the service.

- *Terminal Operator Agreement* (TOA). This agreement is quite different from the previously described since it is much more focused on the business model used by the terminal owner and the terminal operator. It is the most comprehensive of the agreements in this context and builds on the concession process that leads to the signing of this agreement. The usual components of the agreement are fee-structures (e.g. fee per handled container, trailer, swap body, storage of load units, etc.), KPIs, marketing strategy, strategic goals and visions, principles and conditions for infrastructure investments by the terminal operator, time-periods and possible extensions, handover procedures and so on. Another important and often forgotten aspect is the traffic planning and management responsibility of the terminal operator. It is often forgotten because in the early stages it seems an unlikely issue, given that the terminal is newly developed.
- *Terminal Access Agreement* (TAA). This agreement focuses on the routines and conditions (restrictions and specific guidelines) for using the terminal, including from whom and when capacity should be requested and on what basis is capacity allocated.
- *Rail Operator Agreement* (ROA). This agreement contains the terms of the commercial set-up between the rail operator and the terminal operator. Central to this is a list of services and associated prices, for example handling of containers, storage, reefer connections and so on. If the terminal owner has required a public 'same for all' pricing strategy, the prices are often also given through publicly available information from the terminal operator.

Terminal owners may plan for the landlord model, but because of uncertainties and impairments in their contract, it is quite common that they find themselves involved in daily operations. Given that so many aspects related to operations are included in the agreements, this might not be surprising (e.g. traffic management and maintenance). These shortcomings and uncertainties in contracts and agreements are not only an issue for the individual terminal, but also for the system as a whole since they may affect the safety and economic sustainability of terminals. In this respect, it may be very difficult to fix a contractual framework that is dysfunctional since there might be great incentives for some actors not to change the current agreements. From a transport system perspective, great benefits could be derived from more clear and standardised agreements and contracts. The national transport authority has a natural role in this development.

Another important aspect of terminal governance and operations is the change and handover of operations to a new operator. Again, this is something that is often overlooked but can create substantial difficulties. The handover procedure and conditions should be defined in the concession agreement from the start, and there are many components that should be included, for example equipment, semi-infrastructure (small installation such as security), information technology (IT) and staff. The role of the terminal operator might also represent a challenge if, for example, the actor handing over operations is also a rail operator with a large demand for the terminal services. Such a situation could be delicate and difficult to manage; one that is best managed by a clear and well-defined concession agreement. If there is any disturbance in the circumstances surrounding a change of terminal operator, there is a big risk of market uncertainty, and if a transport operator changes terminals because of this, it might be very difficult to regain the flow of those goods.

INTERMODAL TERMINALS FROM A LIFE CYCLE PERSPECTIVE

Table 6.2 illustrates the intermodal terminal from a life cycle perspective, including the main characteristics, activities and stakeholders for each phase. Table 6.2 shows that there are clear differences in the life cycle of terminals and the strategies that may be useful. It also emphasises the fact that the context of terminals is something dynamic, changing over time and differently for different stakeholders.

While, thus far, this chapter has focused on the intermodal terminal, the economic viability of an intermodal terminal is largely dependent on the competitiveness of intermodal transport. The competitiveness of intermodal transport is determined by many aspects, such as handling charges, asset utilisation, balancing of flows of goods and so on. This is the responsibility of the intermodal transport provider, but understanding the terminal life cycle can clarify inputs for long-term decision making for both private and public actors. Overall, the cost of handling at intermodal terminals is important to the cost-competitiveness of intermodal transport. The cost is highly dependent on volumes handled, equipment used, business model defined in the concession agreement and so on. The price, however, is often quite stable as a result of intense market competition and is normally about €20–30 per handled container and about €25–35 for semi-trailers. To a large extent, the cost components determine the economic profitability of the terminal. This results in a common strategy for all terminals to focus on attracting new volumes since the terminals have a low marginal cost for each additional move. The lowering of prices might attract goods from other modes of transport, but there is also a risk of harmful inter-terminal competition. This may result in more terminals operating than the market requires, and some may be kept open with ongoing public subsidies. Given that terminal development is often pursued based on the argument that it will attract new business and employment to a region, it is tempting for a public actor to protect its local terminal and try to ensure its competitiveness by means of public subsidies.

Table 6.2 Summary of main characteristics of and influences on each phase of the intermodal terminal life cycle

	Planning, funding and development	*Finding an operator*	*Operations and governance*	*Extension strategy*
Length	3–10 years	1–2 years	>10 years	>15 years
Main stakeholders	• Public infrastructure stakeholders (e.g. rail authorities, planners and so on.) • Large shippers • Real estate developers • Terminal operator • Rail operators • Ports	• Public infrastructure owner • Terminal owner (if different from the above) • Terminal operator	• Public infrastructure owner • Terminal owner (if different from the above) • Terminal operator • Rail operators	• Public infrastructure owner • Other public stakeholders (e.g. rail authorities, planners and so on.) • Terminal operator
Main activities undertaken	• Planning • Design • Funding sought • Tendering of construction • Construction	• Designing business and ownership model • Tendering for operator • Designing concession agreement • Contract development	• Continuous improvements • Responding to changes in technology and demand	• Renewed terminal concession • Potential changes in business and ownership model • Potential expansion • Ensuring long-term strategy and control • Potential sale and redevelopment of site for new purpose
Main influences	• Existence and location of market demand • Location of competitors • Best practices in design and terminal handling • Availability of innovation and new technology	• Public policy and subsidy • Market structure related to terminal and rail operations	• Market structure (rapid and fast changes to demand), e.g. demand for multipurpose terminal use. • Technology advances • Competition from other terminals and other modes	• Market structure (declining demand or changes to distribution strategies) • Technology advances • Competition from other terminals and other modes • Demand for land from other sectors (e.g. housing, retail)

Relevant policy and regulatory issues	• Interface between transport administration and infrastructure owner • Govt. policy e.g. modal shift, economic development • Planning system, incl. financial incentives	• Interface between transport administration and infrastructure owner • Rail regulations e.g. tariffs, open access	• Interface between transport administration and infrastructure owner • Rail regulations e.g. tariffs, open access • Government policy changes re other modes (e.g. changing regulations on road haulage)	• Government policy e.g. modal shift, economic development • Planning system, incl. financial incentives • Government. policy changes re other sectors (e.g. land rezoning)
Research agenda	• Lack of best practice related to design • Ongoing research on design of multipurpose terminals	• Lack of best practice related to business models, PPPs • Lack of standardised frameworks for tendering and concessions	• Ongoing research on technology advances • Lack of best practice related to active governance e.g. regulation, contracts	• Lack of best practice related to long-term planning and management of strategic infrastructure

Source: Monios, J., Bergqvist, R., *Intermodal Freight Terminals: A Life Cycle Governance Framework*, Ashgate Publishing, London, 2016.

CONCLUSION

This chapter described the role played by intermodal terminals in the intermodal transport system. Terminal design, layout and operations were discussed as well as governance and operational models. Finally, a life cycle governance perspective was introduced, which allows the planning and management of a terminal over the course of its life, generally several decades. There are many important management and operational aspects that the terminal owner and operator must optimise to provide a cost-effective service, which is an essential part of the full rail-service package that the rail operator provides to its users, ultimately the shippers purchasing the transport service.

REFERENCES

Banverket. (2010). Inriktning för godstransporternas utveckling. v.1, BVStrat 1003, Samhälle och planering.

Bärthel, F., Woxenius, Y. (2004). Developing intermodal transport for small flows over short distances. *Transportation Planning and Technology*. 27 (5): 403–424.

Bergqvist, R. (2007). *Studies in Regional Logistics: The Context of Public-Private Collaboration and Road-Rail Intermodality*, Logistics and Transport Research Group, Department of Business Administration, BAS Publishing, Gothenburg, Sweden.

Bergqvist, R. (2008). Realizing logistics opportunities in a public–private collaborative setting: The story of Skaraborg. *Transport Reviews*. 28 (2): 219–237.

Bergqvist, R. (2009). *Hamnpendlarnas Betydelse för Det Skandinaviska Logistiksystemet*, Handelshögskolan vid Göteborgs universitet, BAS Publishing, Gothenburg, Sweden.

Bergqvist, R. (2012). Hinterland logistics and global supply chains. In Song, D-W., Panayides, P. (eds.), *Maritime Logistics: A Complete Guide to Effective Shipping and Port Management*, pp. 211–230. Kogan Page, London.

Bergqvist, R. (2013). Hinterland transport in Sweden: The context of intermodal terminals and dryports. In Bergqvist, R., Wilmsmeier, G., Cullinane, K. (Eds.), *Dryports:A Global Perspective, Challenges and Developments in Serving Hinterlands*, pp. 13–28. Ashgate Publishing, Farnham, UK.

Bergqvist, R., Falkemark, G., Woxenius, J. (2010). Establishing intermodal terminals. *International Journal of World Review of Intermodal Transportation Research (WRITR)*. 3 (3): 285–302.

Bergqvist, R., Monios, J. (2014). The role of contracts in achieving effective governance of intermodal terminals. *World Review of Intermodal Transportation Research (WRITR)*. 5 (1): 18–38.

Bergqvist, R., Pruth, M. (2006). Developing public-private capabilities in a logistics context: An exploratory case study. *Supply Chain Forum*. 4 (1): 104–114.

Bergqvist, R., Tornberg, J. (2008). Evaluating locations for intermodal transport terminals. *Transportation Planning and Technology*. 31 (4): 465–485.

Höltgen, D. (1995). Terminals, intermodal logistics centres and European infrastructure policy. PhD thesis, University of Cambridge.

Janic, M. (2007). Modelling the full costs of an intermodal and road freight transport network. *Transportation Research Part D: Transport and Environment*. 12 (1): 33–44.

Martí-Henneberg, J. (2013). European integration and national models for railway networks (1840-2010). *Journal of Transport Geography*. 26: 126–138.

Monios, J. (2014). *Institutional Challenges to Intermodal Transport and Logistics*. Ashgate, London.

Monios, J. (2015). Identifying governance relationships between intermodal terminals and logistics platforms. *Transport Reviews*. 35 (6): 767–791.

Monios, J., Bergqvist, R. (2015). Operational constraints on effective governance of intermodal transport. *Research in Transportation Business & Management (RTBM)*. 14 (March): 1–3.

Monios, J., Bergqvist, R. (2016). *Intermodal Freight Terminals: A Life Cycle Governance Framework*. Routledge, Abingdon, UK.

Roso, V. (2009). The dry port concept. PhD thesis, Chalmers University of Technology, Gothenburg, Sweden.

Van der Horst, M. R., De Langen, P. W. (2008). Coordination in hinterland transport-chains: A major challenge for the seaport community. *Maritime Economics & Logistics*. 10 (1–2): 108–129.*Table 6.2* Summary of main characteristics of and influences on each phase of the intermodal terminal life cycle.

World Bank. (2007). *Port Reform Toolkit*, 2nd edition, World Bank, Washington DC.

Chapter 7

The port interface

Jürgen Wilhelm Böse

INTRODUCTION

Since the 1960s, intermodal transport* of sea freight has developed rapidly and dominates the break bulk cargo segment on international shipping routes today. This development is due to the use of International Organization for Standardization (ISO) containers (as 20 ft and 40 ft standard boxes†) for cargo transport in supply chains worldwide. In many seaports, the share of non-containerised break bulk cargo is comparatively small nowadays. For example, about 80% of the general cargo handled at the North Range ports of Rotterdam, Antwerp and Bremerhaven was already containerised in 2005, at the Port of Hamburg it was 96.4% (see Notteboom and Rodrigue, 2008).

Because of the specific physical attributes, the efficient transport, handling and storage of ISO containers require the interaction of several specialized logistics systems as well as sophisticated coordination of the related system equipment. In seaports, corresponding systems are used for discharging and loading of container vessels at dedicated handling facilities, the seaport container terminals.‡ In comparison to conventional break bulk handling, the procurement of equipment for this kind of terminal is quite capital intensive. On the one hand, individual equipment units are comparatively expensive; for example the cost of one ship-to-shore (STS) container crane is €8–10 million. On the other hand, terminal equipment is required in large numbers to simultaneously handle several ships (of partly vast sizes) in adequate time. If a port terminal is not able to meet the market requirements due to poorly organised or ineffective processes, it loses competitiveness and with that its main customers, the container shipping companies.

* 'The movement of goods in one and the same loading unit or road vehicle, which uses successively two or more modes of transport without handling the goods themselves in changing modes' (United Nations, 2001: p. 17).

† Dimensions of ISO standard containers: 20 ft box (6.058 m length, 2.438 m width, 2.591 m height), 40 ft box (12.192 m length, 2.438 m width, 2.591 m height).

‡ Subsequently referred to as *container terminals*, *port terminals* or *terminals*.

In the long term, companies might prefer to use another container terminal at the port (if available) or another port in the region.

Against this backdrop, this chapter focuses on container terminals as highly efficient port interfaces in the global network of containerised maritime freight transport. The primary objectives are the characterisation of common terminal types and the comparative analysis of container flows and associated logistics requirements arising within the main terminal operations areas.

For this purpose, the remainder of this chapter is organised as follows. The first section provides some basics on container terminals, including their structure and operational aspects. Next, the major terminal types are highlighted and described according to their characteristics. Based on this information, a generic terminal example is defined for each type. Container movements within the example terminals are then analysed with reference to the associated logistics requirements. The results are used to compare the expected logistics expenditures of the different terminal types. Finally, the concluding section provides a short summary of the chapter contents as well as conclusions on the impact of differences in logistics identified for container terminals.

TERMINAL STRUCTURE AND OPERATIONAL ASPECTS

In the global network of sea freight transport, container terminals act as the physical interface at ports in the handover of containerised cargo between *maritime* or *maritime* and *inland transport systems* (e.g. Rodrigue and Hatch, 2009). As a result of the specialisation of port infrastructure in the last century (Bird, 1963), related facilities are solely dedicated for time- and cost-efficiency container handling. Nowadays, they usually represent one terminal type among others at seaports (see Figure 7.1a).

Depending on the use of technical and human resources, container terminals facilitate three different forms of service integration: They interlink *liner shipping services* with *inland transport services* (inlandServices), liner shipping services among themselves or both. That said, the physical connection of the transport systems involved leads to differing container flows through the terminal:

- In the case of container exchanges between global or regional liner shipping services, the cargo comes from the open sea and, after temporary storage, leaves the terminal again towards the open sea (see Figure 7.1b). Accordingly, the boxes must be double-handled at the quay wall by STS cranes and involve operations processes at the terminal waterside and yard area only. Containers of this type are classified as *transhipment cargo*.
- In the case of container exchanges between liner shipping services and inlandServices, the cargo comes from the open sea and, after temporary storage, departs towards inland destinations (or vice versa). This is associated with operations at the terminal landside as well. Depending on the inland transport system used by the inlandService, boxes access the port hinterland by *road*, *rail* or *inland waterway* and are delivered to consignees or picked

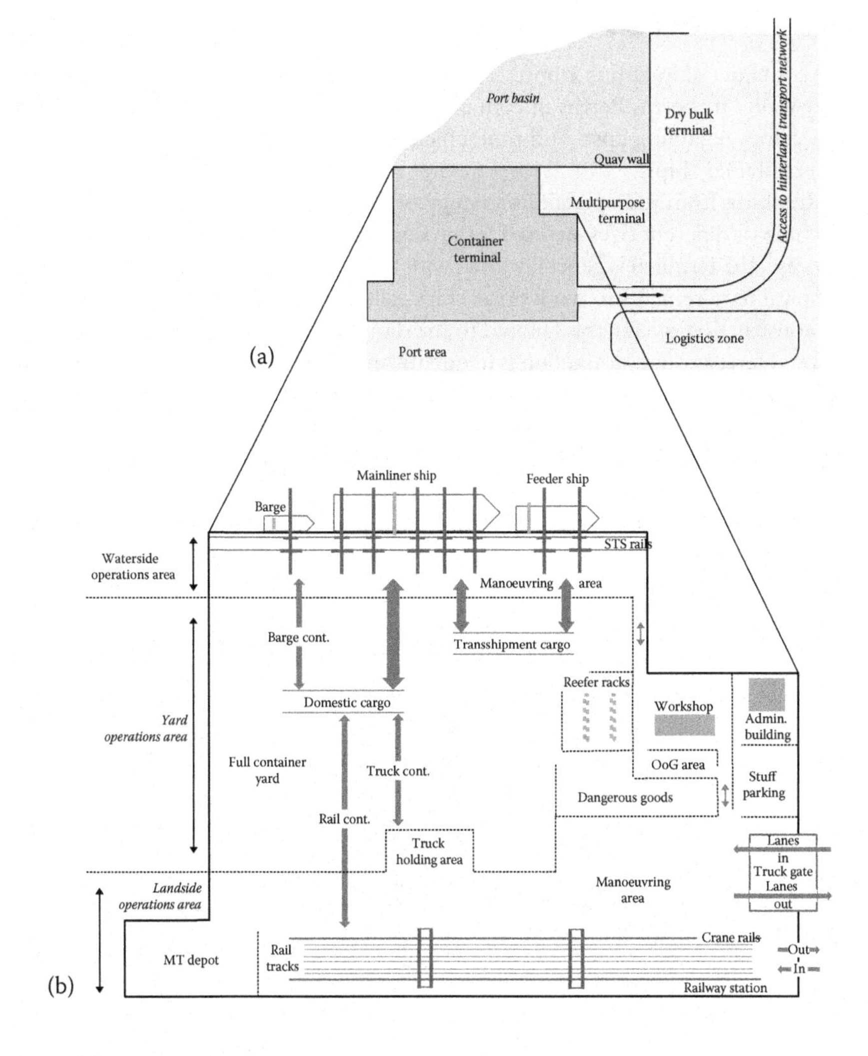

Figure 7.1 Container terminals as integral parts of seaports.

up by shippers. Containers of this type are classified as ***domestic cargo***.

A closer look at the single port terminal reveals a comparatively high complexity with a multitude of internal flow relationships and operation areas, each covering a specific logistics function. In this regard, a basic distinction can be made between the ***waterside***, ***yard*** and ***landside operations area*** of a container terminal. Figure 7.1b shows a stylised illustration of a common container terminal at seaports, highlighting the typical logistical structures. A more detailed description of the main operations areas and their flow-related interdependencies is provided later in this chapter (Böse, 2011; Brinkmann, 2011).

Waterside operations

If a container ship enters a port, it is assigned a certain position at the terminal quay wall – its berth. Berths of container terminals are equipped with specialised STS cranes, enabling cost- and time-efficient discharging and loading of (seagoing) container ships.

Resulting from the operations strategy of the single terminal, quay cranes with the same or different sizes are used for processing ships at the quay wall. In the former case, the terminal is generally fitted with STS cranes of larger sizes, fulfilling the technical requirements to work on all ships calling at the facility. In the latter case, the construction of cranes is tailored to the dimensions of the ships that they operate on. Here, a rough distinction is frequently made between STS cranes* designed for the efficient processing of feeder ships, on the one hand, and (larger) mainliner ships, on the other. The differences in ship dimensions can be seen in Figure 7.2, which shows the feeder *Helmut* with a capacity of 585 TEU† and the mainliner *Maersk Edmonton* with a capacity of 9.082 TEU† at the Port of Gothenburg.

If ports additionally have a high share of barge calls, some of them invest in portal cranes optimised for work on inland ships (see Figure 7.3b). It should

Figure 7.2 Feeder and mainliner ship waiting for processing and in process of loading, respectively, at the quay of the APM Terminals Gothenburg. (From Port of Gothenburg.)

* Variations in STS construction especially concern the *beam length* (over water) and the *lifting height* (over crane rail). For example, STS cranes at the Altenwerder (Hamburg) container terminal differ in parameters, as follows:

Crane use:	*Mainliner*	*Feeder*
Beam length:	61.0 m	28.5 m
Lifting over crane rail:	23.5 m	19.8 m

† Assuming a homogeneous Twenty-foot Equivalent Unit (TEU) at 14 t/TEU.

be noted that many characteristics of barge cranes basically differ from those of classic STS container cranes. The main differences in operating mode relate to the use of the crane portal. While this area for STS cranes is commonly dedicated to container handover with transport equipment and/or out of gauge (OoG) handling, barge cranes typically harness their portal for storage purposes, making a yard area in the back of the cranes superfluous in most instances. Moreover, the main differences in construction between both crane types are evident (see Figure 7.3a and b).

Although some STS cranes also offer the possibility of container handover in their backreach, even the larger ones usually provide not more than eight handover positions for 20 ft containers at maximum. Considering these space

Figure 7.3 Use of STS cranes (a) and barge cranes (From WCN Publishing, 2013) (b) for ship processing.

Figure 7.4 Full container yard and MT container depot at APM Terminals Gothenburg. (From Port of Gothenburg.)

limitations, the comparatively high STS operating frequency* and the large number of containers to be discharged and loaded per vessel, it is not surprising that the crane portal represents one of the main bottlenecks in the waterside handling process (see Figure 7.3a). Thus, on-time container delivery and pick up at STS handover positions is a priority for operations management, to avoid downtime of cranes and, as a result, a loss in berth productivity.

For horizontal transport from and to STSs, different types of terminal equipment are standard today (see Chapter 2).† Such vehicles are specially designed for container movements over longer distances, also possessing (in case of e.g.

* In case of appropriate operating conditions (e.g. related to ship size and the number of transport equipment in use), an average of about 40 container moves per hour during a shift is nowadays possible for modern STS.

† The following vehicle types for horizontal container transport are in operation at terminals today: ***straddle carriers***, ***tractor trailer units***, ***automated guided vehicles*** and ***reach stackers***. For detailed information see, for example, Cargotec (2016) or Terex (2016).

straddle carriers) vertical transport capabilities for container lifting and lowering (see Figure 7.4a). In practice, one vehicle type usually proves to be most beneficial for operating STS cranes, considering local economic and operational conditions.

Yard operations

The main container yard is geographically located at the centre of port terminals, thus accommodating the majority of full containers that pass through the terminal (see Figure 7.1b). Either specialised yard cranes* or vehicles providing horizontal and vertical transport capabilities are used for the handling and stacking of full boxes, (see Figure 7.4a). Due to existing interdependencies between both systems, equipment choices for yard operations and for transport operations at the waterside are frequently made together, based on the achievement of terminal objectives expected from system combinations.†

Furthermore, containers with additional logistics and safety requirements must also be stored at the port terminal. These boxes are moved to or delivered from special storage areas on the terminal site that are partly separated from the actual full container yard (see Figure 7.1b). The main types of related containers (and cargo) are outlined as follows:

- *Reefer containers* require power sockets for energy supply. They are stored by means of steel racks usually providing one socket for two yard slots‡ on several container stacking levels. Moreover, reefer racks enable access to each box for terminal staff. The reefer mechanics in charge monitor, inter alia, the container temperature and manually connect (or disconnect) containers with (from) the electricity grid after (before) their delivery (pick up) at the rack position.
- Containers with *dangerous goods* contain hazardous substances which may be highly inflammable, explosive or poisonous, for instance. For the protection of humans and the environment, this storage area is usually limited in

* Two crane types for full container yard operations are generally considered for port terminals today: *rubber-tyred gantry cranes* and *rail-mounted gantry cranes*. For detailed information see, for example, Cargotec (2016) or Terex (2016).

† Typical system combinations of terminal equipment for waterside transport and yard operations are as follows:

- Pure straddle carrier system
- Automated guided vehicles and rail-mounted gantry cranes with perpendicular block orientation to quay
- Tractor trailer units and rubber-tyred gantry cranes
- Pure reach stacker system (exclusively in use at small port terminals)

‡ One *slot* within a container yard offers space for storing one 20 ft container (i.e. 1 TEU). The slots at the bottom of the yard area (providing space for the first container tier) are termed as *ground slots*.

stacking height and equipped with specific safety devices, such as like a collection basin for leaking liquids or a firefighting pond.

- Analogous to full containers, the large majority of *empty* (MT) *containers* are ISO standard boxes. The maximum stacking height of MT containers is commonly six to eight boxes. By comparison, full containers are merely stacked in piles up to four or five boxes, in many instances. Accordingly, MT container handling requires specific stacking equipment* and is performed within a separate MT container depot differing from the full container yard in its layout structure as well (see Figure 7.4b). Depending on the distance from the depot to the quay wall, as well as given time requirements, it is not unusual that MT containers have a stopover in the full container yard when they are moved from or to the quay wall. In case of loaded containers, this measure is also termed '*pre-stowage*', facilitating or even accelerating the actual ship loading process.
- *OoG cargo* does not fit into ISO containers due to its dimensions or oversize, respectively. Compared with standard boxes, OoG cargo is not stackable and claims more space in at least one dimension, which usually leads to specific logistics requirements.† Thus, OoG cargo is not stored in the full container yard but in a dedicated terminal area aligned to related requirements, for example, with good accessibility from both the terminal truck gate and the STS cranes at the quay wall (see Figure 7.1b).

Finally, it should be pointed out that the 'buffer function' of the terminal yard enables the synchronisation of incoming and outgoing container flows, representing a prerequisite for the smooth integration of all transport processes involved.

Landside operations

The inflow and outflow of containers by *truck* or *railway* are associated with a variety of logistics activities within the dedicated landside operations areas (see Figure 7.1b). The handling of rail containers takes place at a terminal *railyard* equipped with specific railway cranes. For container movements between the railyard and the (different) storage areas, the same vehicle types are considered as those used for horizontal transport at the terminal waterside.

* For the handling and stacking of MT containers, terminal equipment is used providing both horizontal and vertical transport capabilities. Typical equipment examples in this regard are *reach stackers* and *MT container handlers*. The latter represent conventional forklifts equipped with a container spreader.

† For *OoG transport*, tractor trailer units are frequently the choice. Reach stackers or mobile cranes are used for *OoG handling* within the yard area.

Inbound and outbound truck containers are checked at the truck gate for possible damage and their accompanying documents. Depending on the yard system, the actual handling process of the containers is carried out within or at the edge of the main yard using defined handover positions or an indicated truck holding area (see Figure 7.4a). Trucks delivering and picking up OoG cargo or MT containers are directly un-/loaded in the dedicated storage areas.

Further internal container movements are not considered here. Examples in this regard are housekeeping activities in the main yard for consolidating container piles, transport of boxes for pre-stowage purposes or container movements due to repair requirements or customs clearance.

TERMINAL TYPES

This section starts with a classification of major types of container terminals at seaports. To this end, the kind of transport services that a port terminal uses and the direction of container flows through the facility are used as classification criteria.

As liner shipping services form the basis for international maritime container transport, this section distinguishes between two different service types: global *mainliner services* (mainServices), covering container transport between port terminals in different regions of the world, and *feeder line services* (feederServices), moving containers between terminals called by mainServices and other ports of a region, without direct integration into trans-regional service networks.

In the second part of this section, fictitious flow data are used to define an example terminal for each terminal type introduced earlier. All example terminals are characterised by the same *original container flow*, forming the basis for a comparative analysis of the terminal types regarding their characteristic flows and associated logistics requirements.

Terminal classification

Container terminals worldwide do not represent a homogeneous group of facilities, especially considering their function in international transport networks as well as their internal processes. This chapter focuses on the latter, that is the operational aspects and requirements resulting from the container flows of a port terminal. Taking into consideration that facilities are characterised in part by considerable differences in their container flows and transport services, we distinguish between three types of container terminals at seaports:

- Pure transhipment terminal (Trans terminal)
- Combined transhipment and gateway terminal (TransGate terminal)
- Pure gateway terminal (Gate terminal)

Table 7.1 Container terminal classification

Characteristic	*Type*		
	Trans Terminal	*TransGate Terminal*	*Gate Terminal*
Container flow	open sea ↔ open sea (transhipment cargo)	open sea ↔ open sea/port hinterland (transhipment and domestic cargo)	open sea ↔ port hinterland (domestic cargo)
Linked transport services	mainService ↔ mainService/ feederService	mainService ↔ mainService/feederService || mainService/ feederService ↔ inlandService	mainService/ feederService ↔ inlandService

Table 7.1 shows the characteristics of each of these terminal types.*

Trans terminal

Trans terminals exclusively deal with transhipment containers and provide no or insufficient access to the port hinterland. The number of domestic containers at related facilities is, in comparison to the transhipment volume, quite small or negligible. Thus, operations are limited to the terminal waterside, including the discharging and loading of ships, the transport of containers between quay wall and yard as well as their temporary storage within the yard area.

Trans terminals form the basis of the trans-regional and regional network structure in global container shipping. They are transfer points for transhipment containers. If Trans terminals provide a 'linking function' for mainServices, they serve as intermediate storage locations between trans-regional services using the terminal as a ***mainliner hub*** for the recombination and consolidation of shiploads. In the case of the interlinking of mainServices and feederServices, Trans terminals represent intermediate storage locations between trans-regional and regional services. Related ***feeder/mainliner*** hubs divide mainliner containers between departing feeder ships and bundle feeder containers for departing mainliner ships (Rodrigue et al., 2016). Finally, it should be emphasised that

* Depending on the purpose of the analysis, other classification approaches for container terminals can be found in the literature as well. For example, Rodrigue and Hatch (2009) only make a rough distinction between 'container sea terminals' (handling domestic containers) and 'intermediate hub terminals' (handling transhipment containers) for introducing the topic of 'port terminals'. In more detail, Ducruet and Notteboom (2012) classify 'intermediate hub terminals' into 'hub/feeder terminals', 'interlining terminals' and 'relay terminals' particularly aiming at their logistical function in liner service networks. Furthermore, a pure transhipment-related classification approach is given by Rodrigue (2016).

Table 7.2 Key figures for selected container ports

Characteristic	*Port*					
	Port of Singapore[a]	*Port of Shenzen*[b]	*Port of Hamburg*[c]	*Port of Los Angeles*[d]	*Port of Klaipeda*[e]	*Port of Riga*[f]
# terminals	7	4	4	8	2	2
# berths[g]	57	58	25	31	6	2
quay length	17.350 m	17.505 m	7.570 m	9.336 m	1.908 m	645 m
# STS cranes	212	175	80	72	9[h]	7[i]
terminal area	700 ha	792 ha	440 ha	684 ha	54 ha	125 ha
mio TEU (2014)[j]	33,87	24,03	9,73	8,33	0,49	0,39
Transhipment share	85% (2013)[k]	50% (2013)[k]	36% (2015)[l]	<10% (to date)[m]	<10% (to date)[m]	<10% (to date)[m]

a PSA Singapore, 2016.
b Zheng and Park, 2016.
c Hamburg Port Authority, 2016.
d Port of Los Angeles, Container, https://www.portoflosangeles.org/, 2016.
e Drungilas, 2015.
f Freeport of Riga Authority, 2009.
g Berth length: about 300m
h Thereof 4 mobile cranes.
i Thereof 1 mobile crane.
j World Shipping Council, Top 50 world container ports, http://www.worldshipping.org/about-the-industry/global-trade/top-50-world-container-ports, 2016.
k Marine Information Service, 2015.
l Hafen Hamburg Marketing, 2016.
m Rodrigue, J.P., The geography of transport systems: Levels of transshipment incidence, https://people.hofstra.edu/geotrans/eng/ch4en/conc4en/transshipment_incidence.html, 2016.

the individual Trans terminal can also facilitate both container transfer between mainService/mainService and between mainService/feederService. An example in this regard is the Port of Singapore and its container terminals, respectively. Table 7.2 shows key figures for the selected container ports.

TransGate terminal

TransGate terminals handle domestic containers with destinations (or origins) in the port hinterland. Both transhipment and domestic cargo account for an appreciable share of the handling volume of TransGate terminals. Domestic boxes are delivered by mainServices and leave the terminal by inlandServices using *truck*, *rail* or *barge* (and vice versa). Delivery and pick up of domestic containers by feederServices play a subordinate role for terminal operations in many regions of the world.

In summary, handling facilities of this type interlink the port's hinterland (using inlandServices) and geographically neighbouring container terminals (using feederServices) with the global mainService transport network. In addition, TransGate terminals may also function as transfer points at a trans-regional level, enabling container exchange between mainServices converging at their locations.

Container terminals at the Port of Hamburg are examples of feeder/mainliner hubs with a predominant share of domestic cargo. They provide access to a very large port hinterland (up to the Ural Mountains) as well as to smaller port terminals located in the Baltic Sea area.

Other TransGate terminals can be found, for example, at the Chinese port of Shenzhen. Details of the characteristics of both container ports are shown in Table 7.2.

Gate terminal

Gate terminals represent 'gateways' and exclusively facilitate the exchange of domestic containers between inlandServices and mainServices or feederServices. They usually perform no transfer function for transhipment cargo. and if they do, the amount is small or insignificant compared with the number of domestic containers. In some cases, boxes are delivered and picked up by feederServices only. Then, the handling facility can be classified as a *regional terminal* since no direct transport connection is provided to container terminals in other regions of the world. Currently, examples of terminals with a regional orientation are in operation in the Baltic Sea area, such as the container handling facilities at the ports of Klaipėda (Lithuania) and Riga (Latvia). In contrast, Gate terminals, with a dominant share of mainliner containers, can be found, for example, on both coasts of the United States. They are located at ports with a large (and economically strong) hinterland and are well connected to the national rail and road network. Examples in this regard are the Gate terminals at the Port of Los Angeles. Key figures for the Los Angeles container terminals are also provided in Table 7.2.

Finally, the previously mentioned port examples are compared with the 50 largest container ports for the year 2014 (see Figure 7.5). The basis for this comparison is the annual handling volume at the ports' quay wall. The related throughput numbers are commonly used to compare different container terminals or ports since they effectively reflect the volume annually passing through the ports.

Considering that the handling volume of ports ranking around the 100th position still ranges between 1.25 and 1.35 million TEU per year, Figure 7.5 shows the impressive worldwide differences in size between container ports. Additionally, it gives an indication of the enormous complexity inherent in the global liner service network of container shipping.

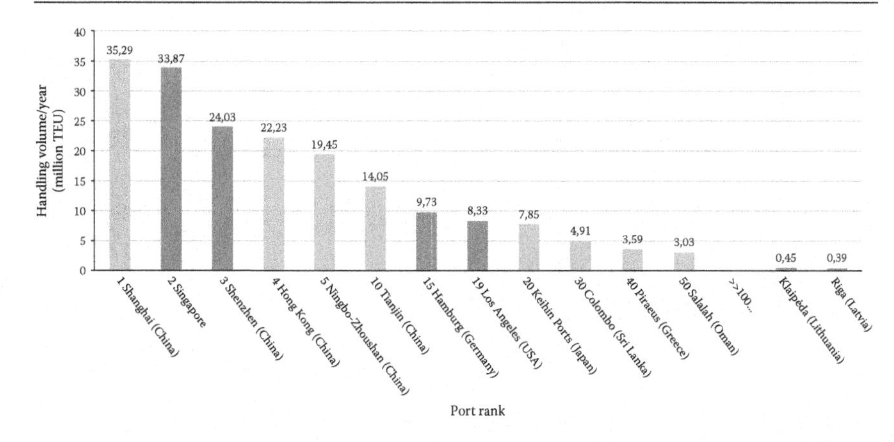

Figure 7.5 Ranking of the largest container ports in 2014. (From author, based on data from World Shipping Council, available at http://www.worldshipping.org/about-the-industry/global-trade/top-50-world-container-ports, 2016, accessed 1 July 2016.)

Example terminals

To analyse the container flows of the already classified port terminal types, three generic terminal examples are defined in this section. The basis for all examples is equal assumptions regarding the *original annual flow volume* of mainliner/feeder containers, counting only the first discharge or loading move of each container – even if the box crosses the quay wall several times. In light of that, the examples are only comparable on this basis if an additional assumption is made – namely, that all containers passing through the facilities are handled at liner ships at least once each time. In other words, a case in which containers enter and leave the terminal by inlandServices is not modelled by the examples.

The original annual container flow of all example terminals amounts to the fictitious quantity of *1 million containers* delivered or picked up by mainliner or feeder ships. Assuming a TEU factor* of 1.6, the corresponding throughput in 20 ft boxes results in 1.6 million TEU per year. Moreover, the liner shares of the original annual flow (mainService vs. feederService), the flow direction (inbound vs. outbound) and the kind of cargo (transhipment vs. domestic) are determined by considering the characteristics of the respective terminal type. Due to the assumptions made here, all other container flows at the example terminals must have a *derivative character*. They are induced by the original flow of containers;

* The TEU factor describes the ratio between the number of containers expressed as *Twenty-foot Equivalent Units* and the actual number of containers, that is, TEU factor = (# TEU) / (# cont.) = (# 20 ft cont. + 2 * # 40 ft cont.) / (# 20 ft cont. + # 40 ft cont.).

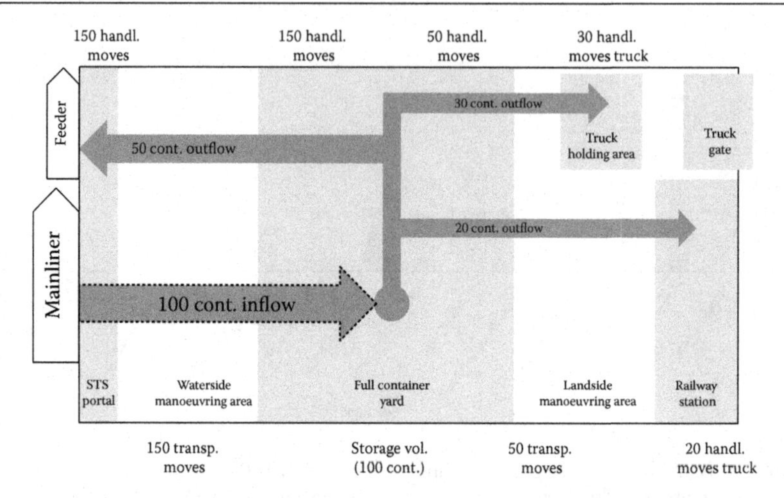

Figure 7.6 **Original main service inflow of 100 containers and resulting derivative outflows with associated logistical requirements (example).**

they represent the upstream or downstream movements of these containers related to their original flow. In this regard, Figure 7.6 presents a simple 'flow example'. Following, for each example terminal, the original and derivative container flows are assumed in line with the underlying terminal type (Figures 7.7 through 7.9).

Trans terminal

For the Trans terminal example, the original annual mainliner volume is chosen to be 850,000 inbound containers, and the original annual feeder volume is 150,000 inbound containers. For the sake of simplification, the feeder inbound share is assumed to be equal to the outbound share, and the exchange volume among feeder ships is set to zero. Accordingly, the original feeder inflow of 150,000 containers induces a (derivative) mainliner outflow of the same size, just as the (derivative) feeder outflow of 150,000 containers results from a corresponding original mainliner inflow (feeder/mainliner transhipment). Appreciable inland container flows do not exist at Trans terminals. Consequently, the remaining original mainliner inflow of 700,000 containers must lead to a (derivative) mainliner outflow of the same size (mainliner/mainliner transhipment) (Figure 7.7).

TransGate terminal

The present example is modelled as a feeder/mainliner hub handling domestic cargo, too. The original annual feeder flow is again assumed to be 150,000 inbound containers, and the original annual mainliner flow is 850,000 containers

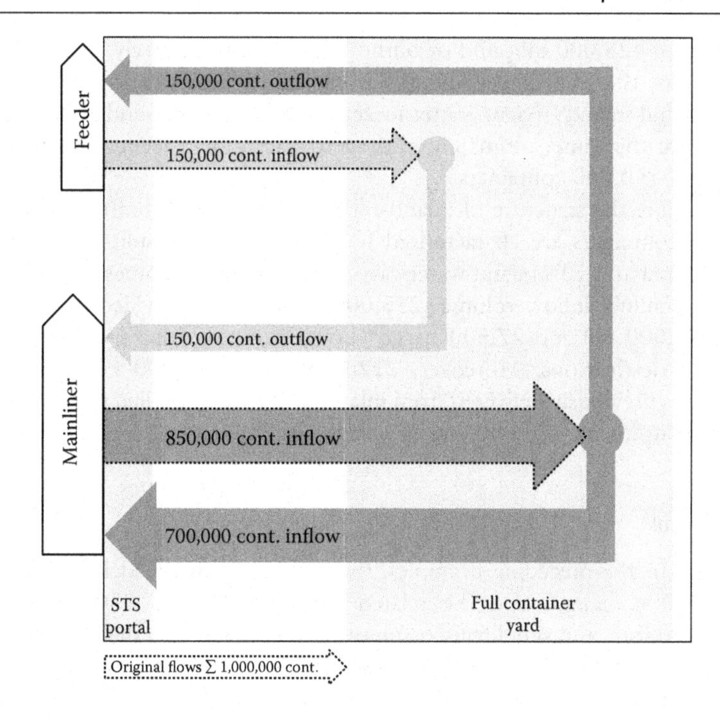

Figure 7.7 Original and derivative annual container flows assumed for the Trans terminal example.

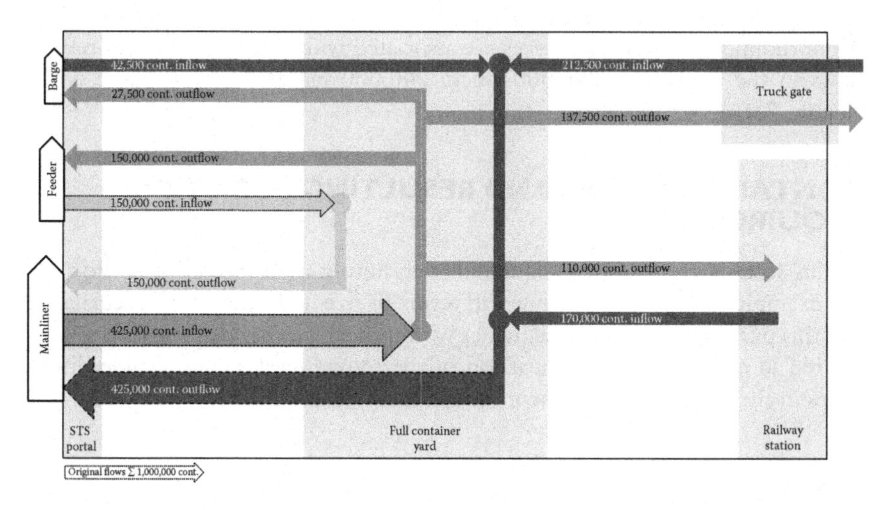

Figure 7.8 Original and derivative annual container flows assumed for the TransGate terminal example.

consisting of 425,000 inbound or outbound boxes, respectively. For simplification reasons, the exchange volumes among feederServices and between feederServices and inlandServices are set to zero. The feeder inbound and outbound volumes are the same, ultimately leading to (derivative) feeder and mainliner outflows of 150,000 containers.

To simplify the structure of inland-related flows as well, both incoming and outgoing containers are characterised by the same modal split – namely 50% road, 40% rail and 10% inland waterway. Accordingly, the domestic share of the original mainliner inflow volume (275,000 containers) is divided into 137,500 truck, 110,000 rail and 27,500 barge* boxes subsequently forwarded to port hinterland destinations. Moreover, 212,500 truck, 170,000 rail and 42,500 barge boxes previously delivered from inland origins are bundled into an original mainliner outflow of 425,000 containers (Figure 7.8).

Gate terminal

Analogous to the preceding examples, the original annual feeder flow amounts to 150,000 containers and the related mainliner flow to 850,000 containers. Furthermore, the simplifying assumptions are made that the original feeder inbound and outbound flows are the same (75,000 containers) just as the original mainliner inbound and outbound flows (425,000 containers). Since Gate terminals provide no liner hub function, derivative feeder/mainliner container outflows do not exist (no transhipment cargo). To divide the original feeder/mainliner inflows into (derivative) inland-related outflows, on the one hand, and to bundle the (derivative) inland-related inflows to original feeder/mainliner outflows, on the other hand, the same 'flow split' is taken into account, as is done with the TransGate example. Accordingly, both the original inbound flow and the original outbound flow of liner services are associated with (derivative) inland-related flows in the amount of 250,000 truck, 200,000 rail and 50,000 barge boxes (Figure 7.9).

CONTAINER FLOWS AND RESULTING LOGISTICS REQUIREMENTS

In this section, the logistical requirements resulting from the container flows of the example terminals are analysed in terms of size and location of occurrence. For this purpose, the logistics activities within the main operations areas are considered in more detail. A comparison of the results reveals considerable differences in the logistical requirements of the example terminals.

* Arriving barges berth at the main quay wall and are operated by the STS cranes actually dedicated to liner ships.

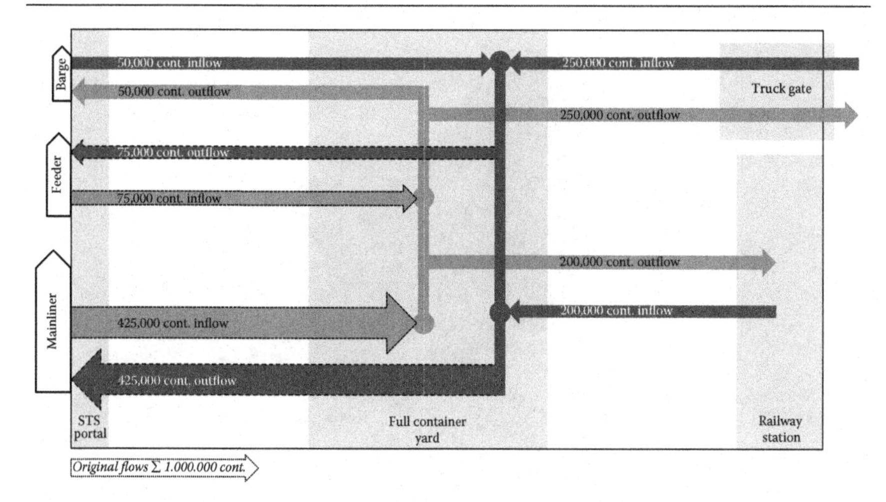

Figure 7.9 Original and derivative annual container flows assumed for the Gate terminal example.

Waterside terminal operations

Logistics activities at the terminal waterside in particular include the discharging and loading of ships as well as the exchange of containers between the quay and yard area. Against this background, Table 7.3 provides an overview of the container movements for all example terminals. Based on the summary of original and derivative container flows to *transhipment* and *domestic cargo*, on the one hand, and *inbound* and *outbound cargo*, on the other hand, the container volumes and the movements induced by them are shown in an aggregated manner. In this regard, it should be noted that the same number of boxes (1 million containers/year) is associated with

Table 7.3 Annual container volume and moves at the waterside of example terminals

	Direction of container moves				
Cargo	*Inbound*	*Outbound*	**Total**	*Share*	*Container volume*
		Trans Terminal			
Transshipment	1.000.000	1.000.000	**2.000.000**	*100%*	1.000.000
domestic	0	0	**0**	*0%*	0
Total	**1.000.000**	**1.000.000**	**2.000.000**		**1.000.000**
		TransGate Terminal			
Transshipment	300.000	300.000	**600.000**	*44%*	300.000
domestic	317.500	452.500	**770.000**	*56%*	700.000
Total	**617.500**	**752.500**	**1.370.000**		**1.000.000**
		Gate Terminal			
Transshipment	0	0	**0**	0%	0
Domestic	550.000	550.000	**1.100.000**	100%	1.000.000
Total	**550.000**	**550.000**	**1.100.000**		**1.000.000**

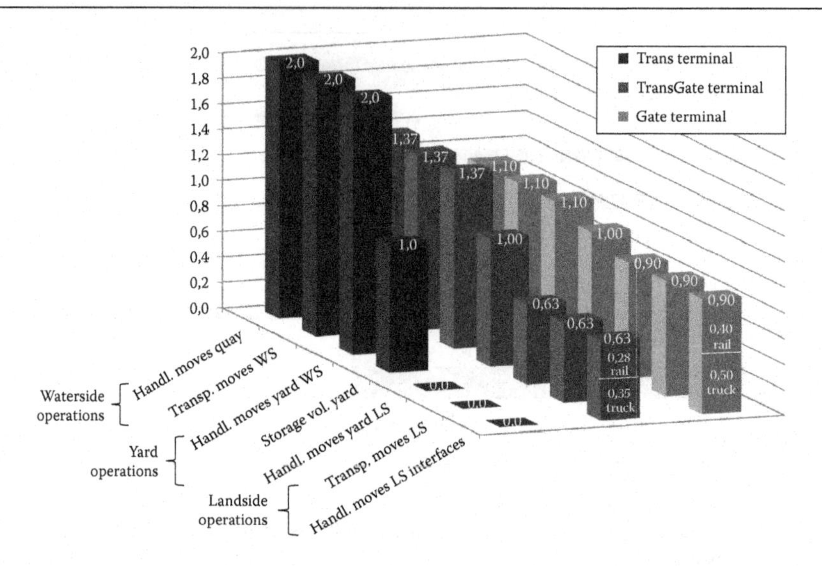

Figure 7.10. **Overview of logistics requirements* for all example terminals differentiated according to terminal operations areas.**

a totally different amount of container moves at the example terminals and thus with correspondingly different effects on terminal logistics. Figure 7.10 shows the transition of related container movements in the logistics requirements expected for the three examples.

The differences in waterside operations are tremendous. While the annual handling/transport workload at the Gate terminal amounts to 1.1 million container moves (based on 1 million liner boxes from which 100,000 are handled a second time at barges), the related workload at the TransGate terminal example accounts for 2 million container handling/transport moves. Due to a 100% transhipment share, in the latter case, each liner box enters and leaves the terminal at the quay wall and thus must be handled (transported) two times.

For the TransGate terminal example, the waterside handling (and transport) workload is mainly determined by the share of transhipment and barge containers because of the double handling requirements of related boxes at the quay

* The numbers in Figure 7.10 indicate the minimum logistics requirements arising from an original annual container flow of 1 million boxes at the example terminals. The requirements can be significantly higher depending on the terminal organisation and the equipment types in use.

Organisation examples: Retrieving moves in the yard are frequently associated with further handling moves of other containers (restacking) or extra transport moves are to be done from and to a terminal near dock site.

Equipment examples: While container handover between an STS crane and an automated guided vehicle requires one handling move (ship↔vehicle), for container handover between an STS crane and a straddle carrier, one STS handling move (ship↔quay wall) and one straddle carrier handling move (pick up/drop off) are necessary.

wall. If the transhipment share of TransGate terminals drops to 0%, the facilities mutate to pure Gate terminals. However, if the transhipment share increases, they increasingly take on the characteristics of Trans terminals.

Yard operations

The logistics activities in the yard of a container terminal particularly concern the storing of incoming containers and the retrieving of outgoing containers. Each box induces at least two handling moves (yard in/out) and additionally requires sufficient space for storing, that is yard slots. With this in mind, Table 7.4 highlights the amount of transhipment and domestic containers, including their inbound and outbound moves at yard waterside and landside, for all example terminals. Regarding these numbers, it is (again) worth noting that the same number of boxes leads to fundamental differences in container movements at the waterside and landside of the example terminals.

Figure 7.10 shows the logistics requirements resulting from the inflow and outflow of yard boxes and the need for storage space. Comparing the three terminal examples, the yard handling workload of the Trans terminal example is completely waterside related. It results from transhipment container transports from and to the STS cranes. By contrast, the yard handling workload at the waterside and landside of the Gate terminal example has a more balanced character and would be equal if 100,000 barge containers did not induce 100,000 extra handling moves per year at the quay wall. In the TransGate example, the related workload is, of course, not equal due to handling both transhipment

Table 7.4 Annual container volume and moves at the yard area of the example terminals

Direction	*Direction of container moves*		
Cargo	*In-/Outbound WS*[a]	*In-/Outbound LS*[a]	*Container volume*
	Trans Terminal		
Transhipment	2.000.000	0	1.000.000
domestic	0	0	0
Total	**2.000.000**	**0**	**1.000.000**
	TransGate Terminal		
Transhipment	600.000	0	300.000
domestic	770.000	630.000	700.000
Total	**1.370.000**	**630.000**	**1.000.000**
	Gate Terminal		
Transshipment	0	0	0
domestic	1.100.000	900.000	1.000.000
Total	**1.100.000**	**900.000**	**1.000.000**

a WS: Waterside, LS: Landside.

and domestic cargo. Basically, the difference in yard handling requirements at TransGate terminals is determined by the amount of transhipment and barge containers.*

Although there are huge imbalances in yard handling requirements, the example terminals show the same need for storage space, assuming identical yard parameters (e.g. container dwell time). The reason for this lies in the equal number of containers annually passing through the yards, or in other words, the same overall size of the original container flow considered for the example terminals (1 million containers/year).

Landside terminal operations

The landside activities of particular logistical importance are the container handling processes at the hinterland facilities' *truck holding area* and *railway station*.† Furthermore, the horizontal transport of containers from and to the facilities is of interest from a logistics perspective.

Table 7.5 shows the annual inbound and outbound flows of road and rail boxes along with the related container volumes. The flow size of the individual inlandService is determined by the modal split of the domestic containers. We note that in contrast to the terminal waterside, each container at the landside induces one move only and that no barge container moves burden the operations

Table 7.5 Annual container volume and moves at the landside of the example terminals

Direction	*Direction of container moves*				*Container*
Cargo	*Inbound LS*[a]	*Outbound LS*	**Total**	*Share*	*volume*
		Trans Terminal			
Truck	0	0	**0**	*0%*	0
Rail	0	0	**0**	*0%*	
Total	**0**	**0**	**0**		**0**
		TransGate Terminal			
Truck	212.500	137.500	**350.000**	*50%*	350.000
Rail	170.000	110.000	**280.000**	*40%*	280.000
Total	**382.500**	**247.500**	**630.000**		**630.000**
		Gate Terminal			
Truck	250.000	250.000	**500.000**	*50%*	500.000
Rail	200.000	200.000	**400.000**	*40%*	400.000
Total	**450.000**	**450.000**	**900.000**		**900.000**

a LS: Landside.

* Assuming barge processing at main quay wall.

† Other hinterland facilities (see Figure 7.1b), such as the 'truck gate' or the 'MT depot' and their logistics processes are not the focus of this chapter.

processes. Regarding the logistics requirements arising from the movements of truck and rail containers, Figure 7.10 provides more details in terms of their type and magnitude. It becomes apparent that the related requirements greatly depend on the transhipment volume of a terminal.

Different from what is assumed for the example terminals, the railyard or other hinterland functions need not necessarily be located on-site but can be represented by a near-dock facility as well. In contrast to on-dock operations, container moves between a near-dock facility and an on-dock facility (such as the container yard or quay wall) require clearance before entering or leaving the terminal area. Rodrigue and Hatch (2009) additionally mention that the travel distance between the near- and on-dock site is not to be considered as decisive for classification but for logistics efforts associated with the related operations. In the view of the port authorities of Long Beach and Los Angeles, for example, near-dock facilities should be located within a 5-mile zone to limit the environmental and economic impact of extra container transport (Parsons, 2006: p. 18).

In the case of near-dock railyards, the site is frequently connected to the actual terminal area by a shuttle transfer service using the terminal's own equipment. Container handover between yard equipment and the vehicles of the shuttle transfer service is then performed on-site at the truck holding area(s). Accordingly, the rail container transports (yard ⟷ holding area ⟷ near-dock railyard) are significantly longer than those of the truck containers (yard ⟷ holding area) at related terminals. In addition, they represent extra workload for the truck gate due to the necessity of container clearance and, as already mentioned, for the truck holding area(s).

CONCLUSION

This chapter focuses on seaport container terminals as highly efficient port interfaces in the global network of containerised maritime freight transport. Based on the characterisation of widespread terminal types, the chapter provides a description of the main container flows and analyses the logistics requirements associated with container movements. To present the analysis in a comprehensible way, generic terminal examples represent the terminal types of interest.

The analysis results reveal fundamental differences in the logistics requirements of the examples and the terminal types. While the annual handling and transport volume at the waterside and landside of Trans terminals is totally imbalanced (a doubling of the original container flow), the related volumes at Gate terminals correspond more or less to each other. Their ratio ultimately depends on the location of barge processing. At TransGate terminals, the imbalance between the logistical requirements at the waterside and landside is determined by the transhipment share and by the barges (if any). Furthermore, it should be mentioned that no logistical differences in the storage requirements exist between the terminal types under comparable conditions. Accordingly, in the case of the same original annual container flow and yard parameters, their need for storage capacity is the same as well. Figure 7.11 provides a general overview of the annual logistics requirements arising at the waterside and landside of a container terminal depending on the transhipment/domestic share or barge handling volume at the quay, respectively. Furthermore, the example terminals discussed in this chapter are appropriately considered.

For the adequate processing of liner services, the lessons to be learned from the analysis results especially concern the great logistical impact of transhipment cargo. Compared with a domestic box, each transhipment container induces twice the number of handling and transport moves at the waterside. This can lead to extraordinary performance requirements for yard equipment and its interaction with the horizontal transport system. Although in the case of pure transhipment cargo, the storage requirements merely correspond to those of pure domestic cargo, double handling of all containers is necessary at the terminal waterside. Accordingly, for hub terminals, the 'right' choice of yard system or the combination of yard and waterside transport systems nowadays represents a

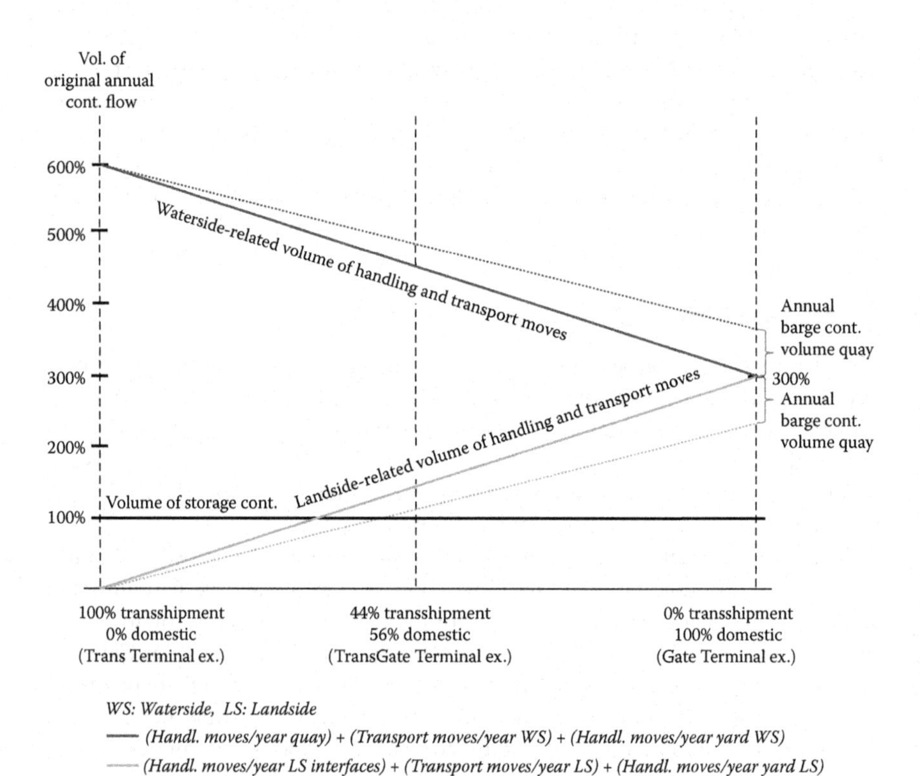

Figure 7.11 Impact on logistics requirements by the transhipment/domestic share and barge processing at quay (minimum requirements).

basic prerequisite to cope with the challenging handling requirements of (main-) liner services.

Regarding the operations of inlandServices, the analysis of example terminals shows that both barge processing at the main quay wall and the use of near-dock facilities can induce considerable extra expenditure and complicate terminal logistics. In the former case, for one thing, the variable share of handling costs per barge container appreciably increases due to expensive STS crane operations. In addition, practical experiences indicate that (despite good planning) barge and liner processing at a common quay wall frequently obstruct each other at TransGate terminals with a higher transhipment share. Here, overlaps in berth and STS crane allocation cannot always be avoided due to the amount of ship calls.

In the 'near-dock' case, the need for transport equipment can significantly exceed the related need of an on-dock facility, depending on the given transport conditions. Furthermore, problems with the container transfer usually arise if a larger quantity of boxes must be delivered at very short notice from the near-dock facility to on-dock sites, such as the container yard, or even worse, to the quay wall. The previous discussion exemplarily shows that the good integration of cost- and throughput-efficient solutions for individual operations areas is decisive for the smooth and successful operation of container terminals as a whole. Both related solutions and their integration, therefore, deserve special attention.

REFERENCES

Bird, J. H. (1963). *The Major Seaports of the United Kingdom*. London, UK: Hutchinson.

Böse, J. W. (2011). General considerations on container terminal planning. In Böse, J. W. (ed.), *Handbook of Terminal Planning*, pp. 3–22. Springer, Berlin.

Brinkmann, B. (2011). Operations systems of container terminals: A compendious overview. In Böse, J. W. (ed.), *Handbook of Terminal Planning*, pp. 25–39. Springer, Berlin.

Cargotec (ed.). (2016). Kalmar: Equipment. Available at https://www.kalmarglobal.com/equipment. Accessed 1 July 2016.

Drungilas, A. (2015). Klaipėda Port—Logistics and Investment Opportunities. Presentation on the Conference on Logistic Sector and Multimodal Transport in Lithuania, 2 October 2015, Klaipėda, Lithuania.

Ducruet, C., Nottteboom, T. (2012). Developing liner service networks in container shipping. In Song, D. W., Panayides, P. (eds.). *Maritime Logistics: A Complete Guide to Effective Shipping and Port Management*, pp. 77–100. Kogan Page, London

Freeport of Riga Authority (ed.). (2009). Freeport of Riga Development Programme—Environmental Report 2009–2018 (technical report). Riga, Latvia.

Hamburg Port Authority (ed.). (2016). Port Information Guide—Hamburg 2016/17 (technical report). Hamburg, Germany.

Hafen Hamburg Marketing (ed.). (2016). Port of Hamburg Press Conference 2016 (presentation slides). 10 February 2016, Hamburg, Germany.

Marine Information Service (ed.). (2015). In Pictures: Top 5 Transhipment Hubs. Port Technology Online News, 25 February 2015.

Notteboom, T., Rodrigue, J. -P. (2008). Containerization, box logistics and global supply chains: The integration of ports and liner shipping networks. *Maritime Economics and Logistics*. 10(1/2): 152–174.

Parsons. (ed.) (2006). San Pedro Bay Ports Rail Study Update (technical report). Irvine, California.

Port of Los Angeles. (2016). Container. Available at https://www.portoflosangeles.org/. Accessed 1 July 2016.

PSA Singapore (ed.). (2016). PSA Singapore Terminals (factsheet). Singapore.

Rodrigue, J.-P. (2016). The geography of transport systems: Levels of transshipment incidence. Available at https://people.hofstra.edu/geotrans/eng/ch4en/conc4en/transshipment_incidence.html. Accessed 1 July 2016.

Rodrigue, J.-P., Hatch, A. (2009). North American Intermodal Transportation: Infrastructure, Capital and Financing Issues, technical report. (Prepared for the Equipment Leasing and Finance Foundation). Washington, DC.

Rodrigue, J-P., Slack, B., Notteboom, T. (2016). The geography of transport systems: Port terminals. Available at https://people.hofstra.edu/geotrans/eng/ch4en/conc4en/centralityintermediacy.html. Accessed 1 July 2016.

Terex. (2016). Terex/Gottwald: Port solutions. Available at https://www.terex.com/port-solutions/en/. Accessed: 1 July 2016.

United Nations. (2001). Terminology on Combined Transport (technical report). United Nations, New York.

WCN Publishing (ed.). (2013). Rotterdam Barges ahead. World Cargo News Online, 20 November 2013.

World Shipping Council. (2016). Top 50 world container ports. Available at http://www.worldshipping.org/about-the-industry/global-trade/top-50-world-container-ports. Accessed 1 July 2016.

Zheng, X.B., Park, N.K. (2016). A study on the efficiency of container terminals in Korea and China. *The Asian Journal Journal of Shipping and Logistics*. 32(4): 213–220.

Part III

FRAMEWORKS

Chapter 8

Intermodal system management and economics

Jason Monios and Johan Woxenius

INTRODUCTION

Chapter 3 covered rail freight transport in detail; therefore, while there will be some overlap, this chapter will focus more on the design of specifically the intermodal system, as well as the role played by intermodal terminals. The following section outlines the design of intermodal transport networks and their key characteristics, introducing the specific challenges of this market and the building blocks that must be managed by operators and other stakeholders. This is then taken further by describing different intermodal transport business models based on varying levels of integration between key stakeholders. The next section sets this information within a theoretical economic context and discusses cost and pricing. Once both the features of the intermodal system and the challenges have been raised, the options available to system managers to address them are considered, such as price incentives, yield management and market segmentation.

DESIGN OF INTERMODAL TRANSPORT NETWORKS

To the shipper, the material flow seems to be direct from the sender to the recipient. But in reality, the directness of transport services depends on the economic and practical viability of consolidating consignments to use the transport resources efficiently. The phenomenon is also referred to as *bundling* and terms used in the rail freight sector include *shunting*, *marshalling*, *classification*, *grouping* and *blocking*. Whether to consolidate or transport directly is decided based on a number of parameters (Woxenius, 2007b):

- Consignment size: The closer to the full capacity of vehicles and vessels, the more direct
- Transport distance: The shorter, the more direct
- Transport time demand: The more specific, the more direct
- Cargo characteristics: The more specific, the more direct
- Availability of other goods along the route: The less the availability, the more direct

When it is decided to consolidate flows, it is generally done in a systematic way using a transport network design. Each network design has inherent pros and cons and matches different preconditions in terms of the character of the transport demand, geography, demography and supply of infrastructure. The feasibility of each network design also depends on when the transport demand is known. If correct information is captured early, there is time to adapt the resources to actual demand, otherwise the operator has to plan according to a prognosis, apply heuristics and choose a rule-based network design. Often, it also means operating with surplus capacity.

Six alternative transport network designs are presented in Figure 8.1. The example is based on 10 nodes with different options of connecting the origin (O) to the destination (D) through transport links. It is assumed that all nodes can be connected with direct links and that all nodes are capable of serving as origins and destinations as well as transfer points.

In the *direct link* alternative, transport is obviously direct from O to D, and there is no coordination with transport between other O–D pairs. The transport *corridor* is a design based on using a high-density flow along an artery and short capillary services to nodes off the corridor. The nodes are thus hierarchically ordered as corridor and satellite nodes. In this example, O is a satellite node and D is a corridor node. In the *hub-and-spoke* layout, one node is designated the hub, and all consignments pass this node, even for transport between adjacent origins and destinations. While the operations follow simple principles, the challenge is to coordinate a large number of interdependent transport services. The *connected hubs* design is another hierarchical layout in which local flows are collected at hubs that in turn are connected to hubs in other regions. It can thus be described as a direct link with regional consolidation.

When using the *static routes* design, the transport operator uses a number of links on a regular basis. In contrast to the hub-and-spoke layout, several nodes

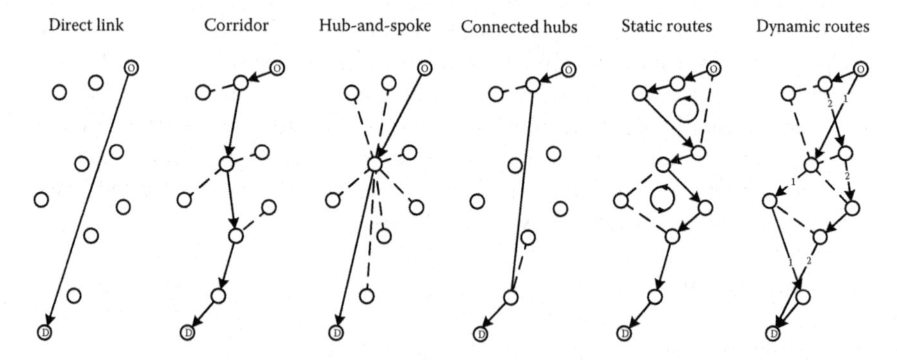

Figure 8.1 Six options for transport from origin (O) to destination (D) in a network of 10 nodes. (Note: Dotted lines show related links in the network designs. In 'Dynamic routes' two alternative routes are shown. In all other designs, the routing is predefined.) (From Woxenius, J., *Transport Reviews*, 27(6), 733–749, 2007.)

are used as transfer points along the route, but transfer might not be needed at every node. Usually, only a part of the load is transferred, and the rest stays on the transport means to the next node. In Figure 8.1, O is on a one-directional loop, connected by a feeder link to a two-directional loop, which in turn is connected to D through another node. The maximum flexibility is offered by the *dynamic routes* design. Consignments are routed through nodes depending on actual demand, and the network operator can choose many different routes between O and D. Transport services are planned by rules of thumb or optimisation methods. In an extreme form, routes can be changed during transport.

Transport networks can be very complex and the layout principles are, of course, not mutually exclusive. One example is domestic hub-and-spoke systems connected to other domestic systems making up a connected hubs system. If the hubs themselves are significant sources, users of a direct link are then combined with users of a hub-and-spoke design and a connected hubs design. Hence, users and operators can perceive networks differently. A passenger or travel agent might perceive most passenger services as static routes, while the transport operators define their services as any of the other designs except dynamic routes.

Consolidation is used in passenger as well as freight transport and there are frequent examples of applications as shown in Table 8.1, also including examples in rail freight transport.

Unit loads are consolidated into train loads in intermodal transport, but each unit load might contain many consignments. The intermodal transport system is then a part of a wider, hierarchical, consolidation network. This section, however, focuses on the traffic designs used for moving unit loads from the origin terminal to the destination terminal by train, hence leaving the pre- and post-haulage by road out of scope.

The network design obviously affects the terminals, both regarding capacity and the time available for transhipment as elaborated by Woxenius (2007b). The *direct link* design is by far the most common and the terminals are either the origin or the destination. All unit loads are transhipped but goods volume handled at the terminals is comparatively limited, thus reducing capacity requirements for the terminals. The transfer time requirements depend on how long the trains are available for handling. If the trains move overnight and stay at the terminal throughout the day, as is fairly common in Europe due to the priority of passenger trains on the shared network, this becomes a non-critical parameter. Due to the customers' timing preferences, however, the terminals are mainly busy in the early morning and the early evening. During those hours, rather fast transhipment is needed and the used transhipment technology must be reliable. The same is true also when the train is used as a shuttle with a tight timetable or if it is used for additional short services during the day. As the fixed costs of expensive rail assets represent the largest share of rail operation costs, using an otherwise idle locomotive or train set for a short run during the day remains an attractive proposition and is a key reason that short-distance intermodal services can be feasible (Bärthel and Woxenius, 2004).

Table 8.1 Typical applications of the different transport network designs in transport services

	Direct link	*Corridor*	*Hub-and-spoke*	*Connected hubs*	*Static routes*	*Dynamic routes*
Passenger	Taxi service	Intercity train service	Domestic airline traffic	Intercontinental airline traffic	Urban public transport systems	Airport limousine service
Freight	Full truck load service	Transport on inland waterways	Air transport of express cargo	Container shipping	Mail and general cargo truck service	Part load truck service
Rail freight	Specialised large-scale solutions	US Class I railroads and short-lines	US double-stack of maritime containers	International wagonload traffic	Classic general cargo service	Old wagonload with frequent shunting operation

Source: Woxenius, J., *Transport Reviews*, 27(6), 733–749, 2007.

To motivate a direct link design, rather large flows are needed to fill the trains, and the services are thus often technically open to several types of unit loads. Semi-trailers are common although the load unit type is somewhat awkward to handle due to the large dimensions and weight. This, in turn, influences the capacity and design of terminals and wagons. Semi-trailers require large and comparatively complicated terminals, and the costs must be distributed between large numbers of annual transhipments. Transhipment is often direct between trains and trucks so storage needs are moderate, but it requires that any unit load can be accessed and transhipped in an arbitrary order. The load plan is important if some prioritised customers are allowed late delivery and early pick-up, when those unit loads should be kept together for efficient operations.

The *corridor* design is sometimes applied in European intermodal transport but has been in large-scale commercial use in Japan for decades. Each train passes several terminals along the route, and the transfer times must be kept extremely short in order not to prolong the total transport time. Nevertheless, only a limited number of unit loads is transferred at each terminal, and it must be possible to operate cost-efficiently on a small scale. The reliability of each terminal is non-crucial since it only affects the specific unit loads to be transhipped at the terminal. The limited distance between terminals also facilitates trucking of unit loads to adjacent terminals in case of breakdown. Since trains are only present at each terminal for a limited period of time, storage space for unit loads must be provided. The requirement of fast transfer and low fixed costs might be contradictory to the transhipment ability of all types of unit loads. Corridor services are thus often limited to a rather homogenous set of unit loads, often implying that semi-trailers or a mixture of container sizes are not accepted. The Japanese system previously mentioned, for instance, is limited to 10-foot containers handled with forklift trucks.

The main characteristic of the hub-and-spoke design is that all unit loads pass through the hub terminal, and it must thus handle a large throughput. It must also be extremely reliable since the whole system is affected if the hub terminal breaks down. The design implies comparatively large detours, and for covering a large area overnight the hub terminal must offer short train stops. Hub terminals can be based on the marshalling of wagons or on transhipping unit loads between trains, thus they are sometimes only designed for rail-rail transhipment and hence are not actually intermodal terminals. In practice, many terminals provide both rail-rail transhipment and regular road-rail intermodal access, focusing more or less on each segment as demand shifts and network design changes. The load plan and exchange technology must facilitate the handling of any unit load, and if all trains combined at the hub are not accessible simultaneously, there is a great need for intermediate storage. The spoke terminals face requirements similar to those of the direct link terminals. If unit loads are only exchanged between a few trains, groups of wagons can be shunted at terminals, requiring that a strict load plan be followed at the spoke terminals, but it facilitates a technically open system accepting a wide variety of unit load types.

Terminals used in the connected hubs design are also either hub terminals or spoke terminals. Fewer trains are connected through the hubs than in a hub-and-spoke design, and consequently the capacity requirements are more modest. Two hub operations consume time, but as the detours are less significant, time requirements for the spoke terminals are similar to hub-and-spoke. The *static routes* and *dynamic routes* designs are not extensively used in intermodal transport and are not elaborated upon here.

INTERMODAL TRANSPORT BUSINESS MODELS

Business models used in intermodal transport are composed of several elements derived from the operational characteristics of the system. The constituent parts are conceptualised in Figure 8.2, which excludes the pre- or post-haul covered in Chapter 5. The figure shows only the terminal to terminal mainline haul. Each terminal may be a road-rail terminal or, for the common port-hinterland transport of maritime containers, one terminal will be an inland road-rail terminal and the other a sea-rail terminal. The latter may be a terminal within the port (known as *on-dock*), whereby the containers are moved only a short distance from the ship to the rail terminal, or outside and nearby the port (known as *near-dock*), in which case the container will be transported by truck through the port gate and perhaps a couple of miles to the road-rail terminal (see Chapter 7).

The analysis presented in this section is relevant mostly to Europe; in the United States, a vertically integrated business model is used, whereby all items in Figure 8.2 (1–6) are operated by a single rail company, who then sells cargo space (5) either directly to a shipper or to an intermediary, whether a third-party logistics providers (3PL), freight forwarder or even a shipping line who offers a door-to-door price to the shipper. Organising the transport of maritime

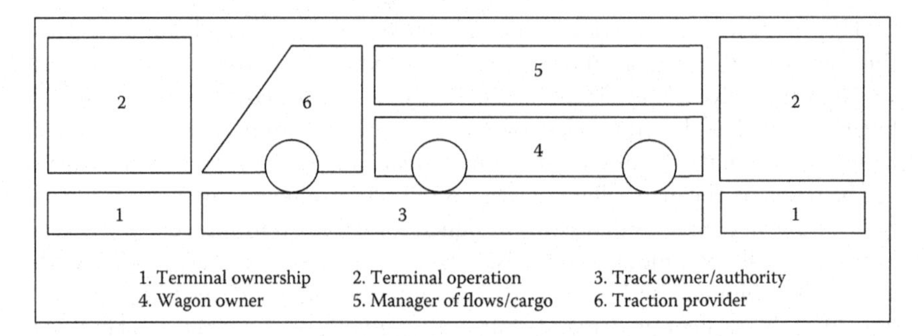

Figure 8.2 Main actors in the rail element of intermodal transport. (From Monios, J., *The Routledge Handbook of Transport Economics*, Routledge, Abingdon, UK, 2017.)

containers between the port and hinterland is divided into carrier haulage, which is when the shipping line books the space with the rail operator as part of the door-to-door service to the shipper, and merchant haulage, which is when the shipper organises their own hinterland transport after the container leaves the ship. Carrier haulage creates more opportunities for efficiencies such as triangulating transport legs, for example transporting a container that has been emptied by an importer to a nearby exporter requiring a container. Merchant haulage requires that the shipper returns the empty container to the port or nominated depot which may result in increased empty transport distance in the system.

It is important to understand the relation between the operational components in the diagram and the management perspective through which they are related. Thus, the definition and role of the intermodal operator may be different under a range of business models. The cargo flows may be managed by an intermodal operator who has their own traction and wagons and sells cargo space directly to the shipper (linking 4, 5 and 6), or the intermodal operator may sell the cargo space to a 3PL or freight forwarder who then sells full container loads (FCL) or less-than-container loads (LCL) to the shipper. If the latter, the 3PL or forwarder will group together consignments from several shippers (known as *groupage*) to produce a full container. The 3PL may also book the entire cargo space on a regular train service at a low rate and then take the risk of selling all these slots to their individual clients. Similarly, the intermodal operator may not operate the traction (6) themselves and may sub-contract a rail operator to provide the traction and/or the wagons. The key aspects are who provides the rail operations, who takes the risk of underwriting the cargo space and how costs and revenue are distributed among actors.

Asset utilisation, or keeping trains full and moving, is essential to achieve economic viability in what is a low margin business, thus the ability of intermediaries such as 3PLs to consolidate flows from many clients is often crucial. One of the reasons why port shuttles are the most successful intermodal venture is because shipping lines with large demand can book several full shuttles in advance because they know that they will have the demand to fill them. Locomotives and wagons are also frequently leased by operators from specialist lessors so that operators can then be more flexible with their provision and not take on too many fixed costs by purchasing too much equipment. Hence, entry barriers are lowered and fixed costs are changed into variable costs. It is also common in the shipping sector for shipping lines to charter some of their vessels for the same reason and only own sufficient vessels to support their core and secure business. While this enables some flexibility in the market, locomotives and wagons are rather specialist equipment so it is not as flexible as a road haulier where trucks are far more interchangeable. In most cases, the track itself (outside the United States) is owned by the public sector and the rail operator will purchase track access for the timetabled routes required.

The situation becomes more complex when considering the relationship between the foregoing business models and the terminal ownership and/or operation (see Chapter 6 for more discussion of terminals). The internal business model relates to the ownership and operation and service provision of the terminal itself, whereas the external business model concerns the levels of ownership, integration and collaboration between the terminal and the external stakeholders such as rail operators, port terminal operators and shipping lines (Monios, 2015). It is possible to have a fully integrated chain where the same company operates the shipping line, the port terminal, the rail service and the intermodal terminal, providing the shipper with a complete door-to-door service, although this is rare. In practice, there is usually a range of levels of collaboration, information sharing and planning between the different actors in each situation.

ECONOMICS OF INTERMODAL TRANSPORT

If road haulage is considered to be rather close to a real-life model of perfect competition between substitutable providers with low barriers to entry and good-quality information, the rail market is quite different. However, the intermodal terminal market is different again to the market for rail operations due to their fixed locations. While the existence of economies of scale in infrastructure is generally accepted, opinion is divided as to whether it exists in operations (Cowie, 2010). The separation of track and operations in Europe was based on the view that economies of scale were limited in operations; therefore, the policy prescription was that it was better to try to obtain the benefits of competition from different operators competing with each other on a shared track. Some evidence does exist, however (Cowie, 1999), that suggests that scale economies do exist in operations, and a number of regional monopolies may in fact be a better model. The US model of competition between a limited number of vertically integrated companies continues to perform successfully, and China is based on vertically integrated regional monopolies, so there is certainly a range of options. Having said that, geographical regions with high demand, long distances and the ability to run long (especially double-stacked) trains produce more scale economies than a fragmented market and geographical scale such as found in Europe where intermodal transport competes not just with road but also with inland waterways and short sea shipping. Intermodal terminals are a combination of infrastructure and services, and different business models exist that attempt to blend the ownership and operation of terminals with the operation of rail services, generally seeking economies of scale and scope through vertical integration (as noted in the previous section).

Mainstream economic theory identifies the key elements that lead to monopoly situations as barriers to entry, minimum firm size, brand loyalty, ownership of key inputs and the potential downsides to society, namely underproducing and overcharging. Barriers to entry are significant in the intermodal terminal market due to high upfront and sunk costs, skill requirements and a lack of terminal locations. These require a minimum firm size, although it may be difficult to put a figure on that size due to high variability in the market between local operators and global operators with multiple terminals. Fierce competition will be faced from incumbents, whether through brand loyalty or aggressive marketing and pricing tactics. This is partly due to their ownership of key inputs, because operating one terminal is only one part of a successful intermodal operation. The terminal must be served by rail services which in turn must connect with other terminals. While all of these indicators of monopoly are present, it does not seem that overcharging is common in the intermodal sector, although data limitations make it difficult for quantitative analysis.

On the one hand, intermodal terminals may be considered a rather substitutable resource as there is not much difference in the basic attributes from one terminal to another. Therefore, like most transport operations, they compete on price subject to an assumed unchanged minimum service quality. On the other hand, they also compete on quality by offering tailored services, which nonetheless must be offered at the same unchanged maximum price that the market will bear. Perhaps this explains why intermodal transport is such a difficult market to grow, demonstrating both the low profit of perfect competition with the underproduction (or inability to release latent demand) and low investment of a monopoly (Monios, 2017). Operators tend to sweat the fixed assets and find it difficult to justify large investments in new assets on such small profit margins.

For an intermodal terminal to be successful, regular traffic is required, which generally means a large amount of production, warehousing or consumption nearby with a suitable distance between origin and destination to support regular long-distance trunk hauls where rail or barge is the natural mode. Various break-even distances have been suggested in the literature (usually averaging at around 500 km, although much lower if pre- or end-hauls are not needed, as in port-hinterland transport), but the reality is that it depends on operational considerations as well as differentials in track charges in different countries and other economies that may or may not have been achieved for example through vertical integration. The longer the distance, the more likely that the increased handling costs of changing mode from road to rail will be offset by the cheaper per-unit transport cost (Figure 8.3). Note that the road sections have a steeper incline due to their higher transport cost per mile compared with rail, but road is cheaper over short distances because of the lower fixed costs.

Rail becomes cheaper than road as distance increases, before sea transport becomes the most cost-effective at longer distances. Figure 8.3 illustrates the importance of the handling time and cost to the intermodal cost function. However, this depends on the quality and the capacity of the intermodal infrastructure as well as suitably scheduled services at the right departure and arrival times, without unnecessary delays along the route. It also depends on the total quantity of cargo, as such services will not be economic unless they achieve high utilisation in both directions. For these and other reasons, road haulage still retains a significant proportion of medium- and even long-distance flows. At short distances, road obviously has the advantage in most cases, but it has proved possible to run intermodal services at short distances, if very high volume is achieved, with good timetables allowing quick turnaround and high utilisation of expensive rail assets. Finally, the administration savings from avoiding port congestion can be another reason to choose an intermodal shuttle, which may offset the higher transport cost. That is why any intermodal scheme (terminal or corridor) must have a clear business model, relating both to transport cost savings (assessing the base transport cost as well as loading and capacity utilisation considerations) and logistics cost savings (including assessment of administration, customs clearance, storage and delays).

In rail services (discussed in Chapter 3), the fixed costs of locomotives and wagons account for the majority of start-up costs, while the proportion of total costs from variable operating costs grows as the fixed costs are spread over a larger volume of output. The marginal cost of adding another container is low,

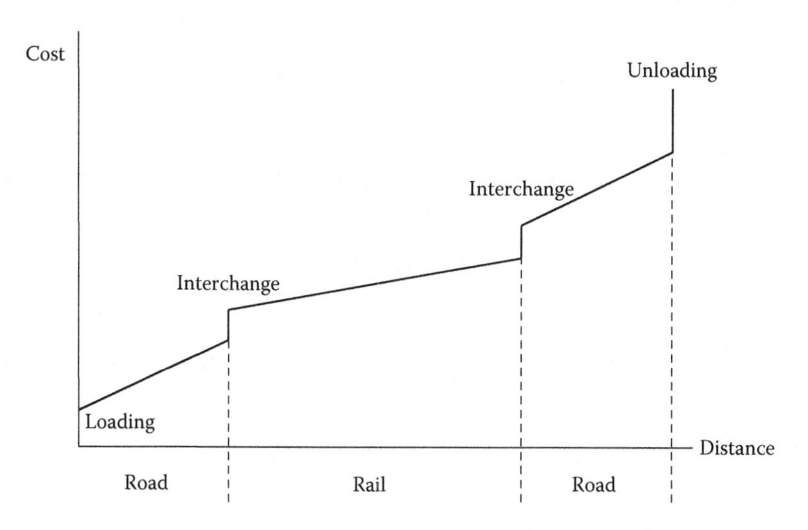

Figure 8.3 Intermodal transport distance and cost. (From Monios, J., *The Routledge Handbook of Transport Economics*, Routledge, Abingdon, UK, 2017.)

except for when a new service needs to be put on, requiring a new payment of fixed costs which is what produces a stepped cost function (Figure 8.4).

It is similar for terminals but with longer steps, because the fixed costs (e.g. terminal area, track, cranes) will only need to be increased very occasionally; therefore, there will be a long period of output growth at marginal cost before another investment step needs to be taken. Once fixed costs have been recouped, keeping variable costs to a minimum is the real challenge for terminals, where unnecessary delays can arise due to disputes over terminal capacity, priority access, maintenance and bad weather events. It is often the case that poorly specified contracts lead to uncertainties and disputes that can produce unforeseen delays and costs (Bergqvist and Monios, 2014).

INTERMODAL SERVICE MANAGEMENT

Intermodal transport needs to operate at high utilisation levels to compete with road and to some extent maritime transport. One of the most important operational goals is to increase the utilisation of rolling stock. The basic idea behind unit load transport is, in fact, to shorten transhipment times to make the transport resources available for transporting other consignments and thus generate more revenue. The current practice whereby trains often stand idle at terminals during the day implies rather bad time utilisation of the rolling stock, yet using train assets for short daytime services faces challenges as they must still align with the dominant practice of overnight freight movements (Bärthel and Woxenius, 2004).

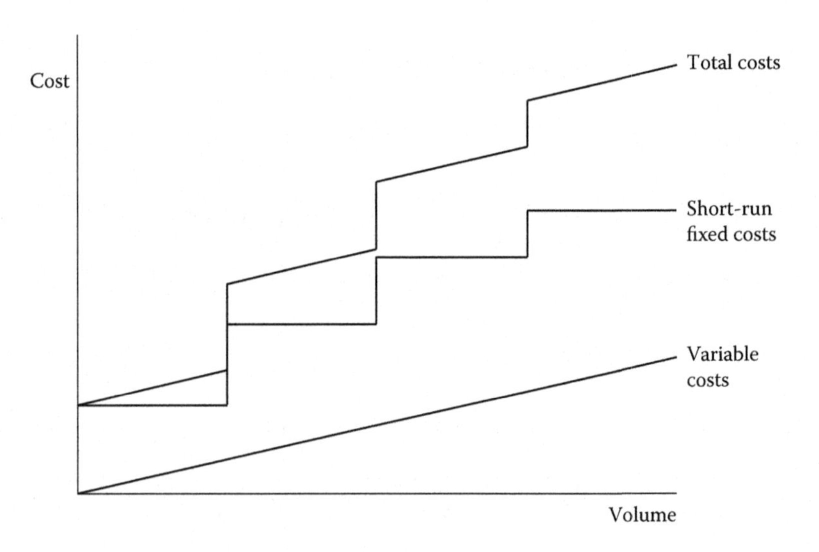

Figure 8.4 Stepped cost function for intermodal transport. (Based on Harris, N.G., *Planning Freight Railways*, A&N Harris, London, UK, 2003.)

Intermodal operators also need to maximise the utilisation level for each train departure. As the majority of rail costs come from fixed asset costs, the marginal costs for a new service are often low as they are mostly the variable costs of fuel and labour. Empty slots on trains are, therefore, a waste of fixed costs and disastrous for competitiveness. This section discusses the options for capacity management at the operational level available to intermodal operators. The measures for increasing the loading space utilisation and thus revenue for intermodal services (Woxenius et al., 2013) include the following:

- Adapting the trains' capacity.
- Adapting the departure timing.
- Using trucks parallel to rail lines, that is, replacing badly filled trains with road transport.
- Adapting train routes, that is, optimising the train routes dynamically to changes in demand.
- Assigning terminals dynamically, that is, optimising the assigned departure and arrival terminals for each load unit.
- Applying business models where shippers or forwarders have to reserve fixed capacity, hence transferring the business risk of running empty slots.
- Applying price incentives, that is, motivating customers to fill empty slots by offering dynamic pricing.

To apply these measures often requires improved information sharing and exchange between the involved actors, including shippers, freight forwarders, road hauliers, terminal operators and rail operators. It is also often necessary to develop and implement decision support systems that rely on operations research methods or agent technology.

Reducing average costs and maximising revenue by a high load factor on trains is only a secondary goal. The operators' main goal is obviously to maximise the business's economic profit. The low marginal costs in rail operations mean a rather straightforward correlation between filled trains and the economic result of the operations. Understanding the marginal costs and the character of the demand is thus critical to managers' ability to improve profitability.

Organisational structure and the scope of the services offered to the market obviously influence the operational decisions on slot utilisation. For the remainder of the chapter, the intermodal system is viewed as a single entity encompassing a set of roles or functions. The assumption that all roles aim at improving the system performance is admittedly a simplification, since there is often an element of opportunistic behaviour in a serially coupled production system.

Adapting the train's capacity to the actual demand or at least the estimated demand is the first measure that addresses loading space utilisation. It regards the time between departures and the number of wagons per train, but it does not regard the track and terminal capacity, since that is regarded as given at the operational decision level. To adapt the capacity, one option is to wait until the train is full before departing, similar to small ferries crossing rivers and narrow straits. Waiting for departure until the train is full is rarely accepted by shippers, and it also implies a risk of insufficient utilisation of the intermodal operator's rolling stock, terminals and the rail infrastructure. To ensure a high load factor, another option is to follow a strict timetable but create a shortage by limiting the number of trains below demand, resembling the attempts by container shipping lines to lay up ships to match capacity to demand. It also implies lost revenues and as this strategy is also unpopular among shippers, it involves a risk of losing the shippers with the highest willingness to pay since they can afford using the all-road alternative. A third option is to apply a fixed frequency but vary the capacity by altering the number of wagons that are available in each train. The marginal hauling costs related to one empty rail wagon are nevertheless low compared with the operational costs of forming trains as well as the planning costs. Splitting trains between departures might, however, be rational if there is a fixed number of wagons to use in a network. European intermodal transport is dominated by shuttles rather than by dynamically composed trains and networks.

The second option is *adapting departure timing*. Although time is usually decisive for the choice between road and intermodal transport, timing is often more important than speed for goods with relatively high value over shorter distances. The principle is still production and sales in the daytime and transport overnight and this is often the only available option since passenger trains occupy the tracks during the day. The processing industry and the container hinterland services do not follow this production cycle implying overnight transport but most other segments demanding long-distance transport do. A large share of the transported goods in line train systems belongs to strictly time-coordinated supply chains. Sommar and Woxenius (2007) demonstrated that if the departure times of trains do not fit the transport chain demands, then the customer in terms of shippers, road hauliers or forwarders simply use trucking instead. Consequently, timing departures according to demand is a major measure for attracting loads to the trains, as further developed in Chapter 9.

The former measure assumes that all long haulage will be performed by rail but there is no end in itself for a transport provider to maximise the use of trains. Certain unit loads might well be more rationally transported by road. A measure focusing on profitability is thus *using trucks parallel to rail lines*. Relevant situations include a too low general load factor of train departures, a severe imbalance or unexpected service interruptions. A specific situation is when a few surplus unit loads remain at a terminal. Lorries can transport these loads and avoid a second train departure with the step in the cost function discussed in the last section. An intermodal operator may also use trucks when it builds up flows to the scale where trains would be economically feasible or as temporary backup during periods when diminishing flows no longer justify train operations.

By using good information exchange and information about the booking situation for a planning period ahead, the train dispatcher can use the measure of *adapting train routes*, and hence apply the traffic principle of dynamic routes, when this dispatcher has alternative paths to use through the network. The dispatcher could route trains differently for each departure to maximise the utilisation of train resources or to maximise revenue. This measure is similar to the operations of part load trucking or buses routed from an airport to hotels depending on the actual demand for travel, hence using the dynamic routes network design as previously discussed. It can also imply avoiding stops with no demand for transhipment and adapting the timetable accordingly to shorten the total transport time. Dynamic routing requires flexible access to track infrastructure via either slack capacity or dummy slots booked in advance. Track access plans are only updated a few times a year, and the dynamic allocation of slots is complicated by an increasing number of operators and passenger trains would still require fixed timetables for the convenience of passengers and would remain sensitive to disturbances.

Dispatchers fine-tuning the demand and supply of slots in an intermodal train following a corridor design may face a short demand overlap, such as a capacity bottleneck between two stops. Slight overbooking is possible on intercity trains when a few passengers can stand waiting for a seat to be free. Overbooking is obviously impossible in a unit load train with a fixed number of slots. It can be solved by denying transport of any of the overlapping unit loads or by re-routing a pre- and post-haulage lorry to a farther terminal. The measure, *assigning terminals dynamically*, requires an efficient two-way information flow, and probably also price incentives for the customer whose unit load has to be routed to another terminal. In an integrated system under one management, as is assumed here, the cost can be absorbed if the marginal cost, including the change of terminal, is lower than the marginal revenue.

Taking a transport planning rather than an inter-organisational perspective, the purpose of *applying price incentives* is to maximise revenue, or in other words, to apply yield management. Yield management is explored extensively in the airline industry, increasingly in passenger train operations but, so far, rarely in rail freight transport. Yet, rail freight satisfies the requirements of yield management, that is, buyers who cannot easily resell the service and thus opportunities for shippers to speculate in transport service contracts. For price incentives to have an effect, some shippers must also be price sensitive. Compared with passengers, shippers typically evaluate more parameters than price and are more often bound by long-term contracts. At least some shippers might anyway consider changing behaviour to benefit from price incentives. The transport provider offers transport services between combinations of origin and destinations along a rail line, which makes the problem different from yield management on a direct link, as is often the case in air transport. The load factor along the line generally varies significantly and the pricing between specific pairs of terminals is then the most obvious base for price incentives. Prices will be increased for services between terminals where there is a significant risk of shortage of capacity and vice versa to attract transport demand between terminals with surplus capacity. This may deter some customers, but ideally enough customers would consider using another terminal to avoid the expensive bottleneck link. Arrival time precision, that is, a wider time window, can also be used for price differentiation as well as offering standby shipments at attractive prices. Then, the intermodal operator can store some unit loads, which can be moved by alternative departures, thereby increasing the total load utilisation. It can also be used for limiting the total capacity (e.g. by using fewer wagons) and still offer the same service level (i.e. the probability of offering the customer a desired service). Examples of factors for deciding the price at any given moment are the competition, the time remaining to departure, the current booking situation and planned or forecasted load utilisation.

Conditions for order time and distribution of obligations between the actors are decided upon above the operational level, yet they clearly influence the operational level. *Improving information sharing* is a prerequisite for several other measures, but it is also a specific operational measure. Information about the transport demand is often available in the synchronised supply chains well before the train departures but that does not mean that it is conveyed to the transport operators if it is not stipulated in the contract. Obviously, this reduces the provider's ability to improve. Operational-level managers then need to inform higher-level managers to negotiate better information sharing.

The foregoing measures require that data is processed into information, transferred and made available to each actor when needed. *Applying a decision support system*, the information is used for developing and sharing efficient plans across several partners in transport chains. Normative methods such as optimisation can be encapsulated in the decision support system. Most of the decisions needed for applying the various measures are taken under uncertainty regarding the demand. The interactions between the measures are rather complex, such as price incentives with capacity allocation or timetables with demand adaptation. Computerised support such as agent technology could be used for operational decisions for prioritising between consignments, such as choosing which unit loads should be left at a terminal when a train is full. Each consignment is then represented by a software agent including the utility function of the consignment, that is, how urgent the delivery of the goods is considered. Then, the modelled agents can even be integrated in the calculation to determine which shipper suffers the least by waiting. This can be done by using a market-based approach like a computational auction. The intermodal operator then applies something similar to yield management, but the customers are represented by software agents rather than the customers' employees. By using micro-level simulation, a transport provider can analyse potential timetable or pricing policies through what-if scenarios.

CONCLUSION

This chapter has outlined and described the key elements of intermodal system design, based on the economics of intermodal transport and the business models through which the system is managed. An understanding of the system reveals that many limitations are placed on efficient operations and constrain the ability of the intermodal system to compete with road. However, a number of options are available to system managers, such as price incentives, yield management and market segmentation. As discussed in Chapter 9, obtaining the best outcome from such practices generally involves a degree of integration between the key actors and a modification of the logistics set-up of the company.

REFERENCES

Bärthel, F., Woxenius, J. (2004). Developing intermodal transport for small flows over short distances. *Journal of Transportation Planning and Technology*. 27 (5): 403–424.

Bergqvist, R., Monios, J. (2014). The role of contracts in achieving effective governance of intermodal terminals. *World Review of Intermodal Transport Research*. 5 (1): 18–38.

Cowie, J. (1999). The technical efficiency of public and private ownership in the rail industry: The case of Swiss private railways. *Journal of Transport Economics and Policy*. 33 (3): 241–252.

Cowie, J. (2010). *The Economics of Transport: A Theoretical and Applied Perspective*. Routledge. Abingdon, UK.

Harris, N. G. (2003). Operational planning. In: N. G. Harris and F. Schmid (Eds) *Planning Freight Railways*, pp. 137–154. A&N Harris. London, UK.

Monios, J. (2015). Identifying governance relationships between intermodal terminals and logistics platforms. *Transport Reviews*. 35 (6): 767–791.

Monios, J. (2017). The economics of intermodal transport. In: J. Cowie and S. Ison (Eds) *The Routledge Handbook of Transport Economics*. Routledge. Abingdon, UK. In press.

Monios, J., Bergqvist, R. (2016). *Intermodal Freight Terminals: A Life Cycle Governance Framework*. Ashgate. London, UK.

Sommar, R., Woxenius, J. (2007). Time perspectives on intermodal transport of consolidated cargo. *European Journal of Transport and Infrastructure Research*. 7 (2): 163–182.

Woxenius, J. (2007a). A generic framework for transport network designs: Applications and treatment in intermodal freight transport literature. *Transport Reviews*. 27 (6): 733–749.

Woxenius, J. (2007b). Intermodal freight transport network designs and their implication for transhipment technologies. *European Transport/Trasporti Europei*. Issue 35, April: 27–45.

Woxenius, J., Persson, J., Davidsson, P. (2013). Utilising more of the loading space in intermodal line trains: Measures and decision support. *Computers in Industry*. 64 (2): 146–154.

Chapter 9

Intermodal logistics

Jason Monios and Rickard Bergqvist

INTRODUCTION

This chapter examines the logistics requirements for modal shift to intermodal transport. Intermodal transport is naturally suited to certain types of flows, based on product and route characteristics and supply chain decisions such as inventory management and storage location. These decisions and strategies are exemplified based on two empirical case examples. The first is of UK retailers using intermodal transport for secondary distribution from centralised national distribution centres (NDCs) to local distribution centres (DCs) and then to stores, and the second is of shipper-forwarder vertical integration in Sweden.

CHALLENGES TO THE ADOPTION OF INTERMODAL TRANSPORT

Modal shift from road to rail faces a number of challenges, which are discussed in more detail in Chapter 3. The customer desires low transit time, reliability, flexibility and safety from damage; while intermodal transport generally performs well in terms of reliability and security, transit time and flexibility present more of a challenge, hence its suitability for certain types of product. Aspects of intermodal transport that limit its ability to compete with road haulage include distance, lack of flexibility, lead time for service development and the role of the last mile (Bärthel and Woxenius, 2004; Slack and Vogt, 2007). In addition, the high fixed costs of rail operators and the requirement to consolidate flows on key routes make profitable service development difficult. As discussed in Chapter 3, setting up a rail service is a complicated task, which is a barrier to intermodal growth and also a barrier to market entry for new rail operators (Slack and Vogt, 2007). In some cases, government grants and subsidies facilitate rail service development, but the availability and type of such funding differ by country and policy mix.

Eng-Larsson and Kohn (2012) found that when making a decision to use intermodal transport, the convenience of the purchase was often more important than the price. Cooperation is needed to achieve economies of scale on certain routes, but research has found industry reluctant to pursue such a strategy

(Van der Horst and de Langen, 2008). Some cases exist of close collaboration between shippers and intermodal service providers and third-party logistics providers (3PLs) (Monios and Bergqvist, 2016; Jensen and Sorkina, 2013) which are often essential in overcoming the fragmentation in transport chains, but many challenges exist that are often only overcome by high levels of trust.

Collaboration in intermodal logistics includes the shippers themselves as well as rail operators and 3PLs who provide a necessary consolidation and integration role. Schmoltzi and Wallenburg (2011) studied horizontal partnerships in logistics and found that while almost 60% of 3PLs in their study operated at least one horizontal partnership with other 3PLs, the failure rate was below 19%, against an average failure rate for horizontal collaborations in many industries ranging from 50% to 70%. The authors also found that, while horizontal collaboration might be thought to be based on cost reduction, the primary motivations revealed in their study were service quality improvement and market share enhancement. Similarly, Hingley et al. (2011) found that cost efficiencies from horizontal collaboration were less important to grocery retailers than retaining supply chain control. These findings have important ramifications for intermodal logistics because it is not simply about reducing transport costs by switching mode but more about integrating intermodal links within their logistics set-up.

Another overlooked issue is that an intermodal service needs to be well developed before shippers will use it (Van Schijndel and Dinwoodie, 2000). Thus, it is important for a third-party consolidator such as a 3PL to amalgamate the loads of several customers to keep a train full and provide a regular service so that other users see this and their confidence in its reliability is increased. The challenge to this view is the evidence of high inertia in the industry when it comes to changing an existing logistics structure. Runhaar and van der Heijden (2005) found that over a proposed 10-year period, even a 50% increase in transport costs would not make producers any more likely to relocate their production or distribution facilities. Woodburn (2003: p. 244) investigated the relationship between supply chain structure and potential for modal shift to rail, and found that 'for rail freight to become a much more serious competitor to road haulage would require considerable restructuring of either the whole logistical operations of companies within supply chains or far-reaching changes to the capabilities of the rail industry to cope with the demands placed upon it'. Therefore, in order to make intermodal transport more effective and attractive to users, it is not only the quality and cost of the service itself that must be considered but also the potential to modify existing transport requirements, which derive from the logistics operation and the broader supply chain decisions of the shippers.

MODIFYING THE SUPPLY CHAIN TO SUIT INTERMODAL TRANSPORT

While the discipline of logistics is a vast realm beyond the scope of this chapter, a brief introduction is required in order to understand how to align intermodal transport constraints with the requirements of a logistics system. There are many definitions of logistics; the one used here is 'Logistics is the process of planning, implementing and controlling the efficient, effective flow and storage of raw materials, in-process inventory, finished goods, services, and related information from point of origin to point of consumption (including inbound, outbound, internal and external movements) for the purpose of conforming to customer requirements' (Coyle et al., 2003: p. 38). Logistics management is sometimes defined based on the seven Rs: getting the right product to the right customer in the right place at the right time in the right condition in the right quantity for the right price. Supply chain management is a broader concept, defined by Christopher (2011: p. 3): 'The management of upstream and downstream relationships with suppliers and customers in order to deliver superior customer value at less cost to the supply chain as a whole'. Thus, transport is one part of logistics which is itself one part of supply chain management.

As discussed in Chapter 8, integration is an important aspect of successful intermodal transport, whereby terminals and transport service operators produce a joint service, sometimes even involving the active participation of the shipper. While the decision on what degree of integration to pursue can be based to some extent on transport decisions, there are also logistics influences such as the location of DCs and the product characteristics that influence the ability to plan services and consolidate loads. Some recent trends of logistics management that influence the potential application of supply chain integration are the centralisation and relocation of plants and DCs, a reduction in the supplier base and a consolidation of the carrier base (Lemoine and Skjoett-Larsen, 2004; Abrahamsson and Brege, 1997). A less fragmented system can reap economies of scale and scope and enable more integrated planning. Supply chains are being reconfigured around the rationalisation of transport requirements, changing distribution strategies and new hub locations. The trade-off between transport and warehouse costs (Figure 9.1) in selecting the optimum number of warehouses has created a centralised structure which has the additional benefit of being suitable for intermodal transport on certain trunk routes between a large NDC or export distribution centre (European distribution centre [EDC]) and a terminal that can then serve local DCs.

The trade-off between transport costs and other costs in order to maintain or reduce total costs is an essential aspect of intermodal logistics. If a customer is going to switch to intermodal transport then they expect that their total costs will either stay the same (but they perhaps benefit in other ways from using intermodal transport, such as for corporate social responsibility) or, ideally, be reduced by transport cost savings. The logistics challenge is that using intermodal transport may increase inventory costs if higher stock levels are needed, which in turn may require larger warehouses, so any decrease in transport cost may be offset by an increase in other cost areas.

Primary distribution refers to inbound flows into the DC. These flows can come from overseas through ports, by rail or air or they can come from within the country. Generally, port flows are non-food lines such as clothes or electronics

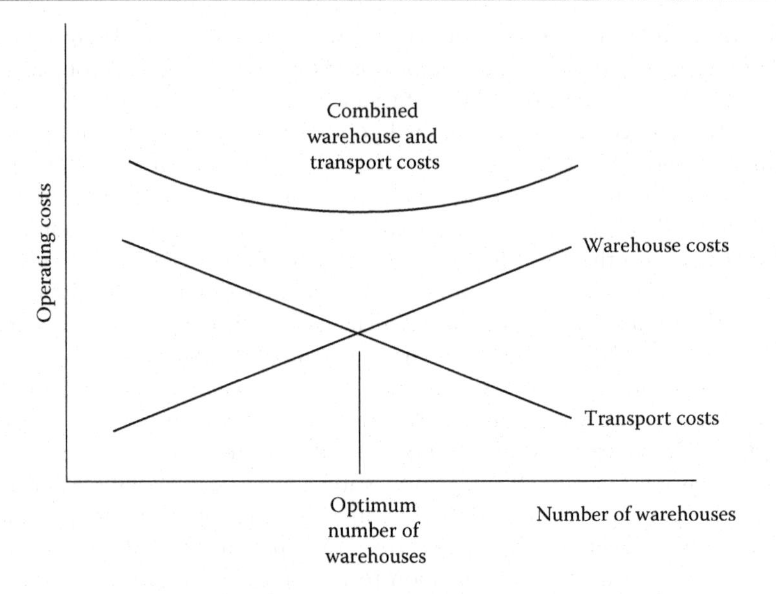

Figure 9.1 Trade-off between transport and inventory costs. (Based on Ballou, R.H., *Basic Business Logistics*, Prentice Hall, Englewood Cliffs, NJ, 1987.)

from the Far East moving through deep-sea ports. These will be consolidated in the NDC. Secondary distribution refers to the movement from the DC to the store. This move is more likely to be done in-house by the shipper or sub-contracted on a closer relationship.

The spatial distribution of the logistics sector, particularly retail, has evolved over the last few decades (Figure 9.2) from a system whereby suppliers delivered directly to stores to the introduction of DCs in the 1970s and 1980s to the arrival

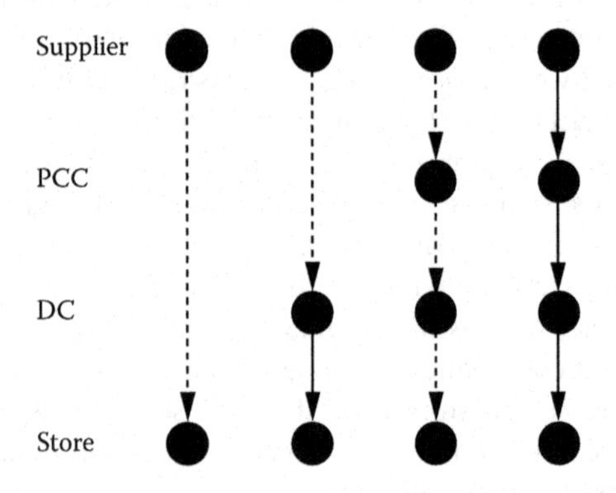

Figure 9.2 Spatial and operational evolution of grocery supply chains in the United Kingdom. (Based on Potter, A. et al., *International Journal of Retail and Distribution Management*, 35(10), 821–834, 2007 and Fernie, J. et al., *International Journal of Logistics Management*, 11(2), 83–90, 2000.)

in the 1990s of primary consolidation centres (PCCs) (Fernie and McKinnon, 1991; Fernie et al., 2000; IGD, 2009). Lead times and inventories were greatly reduced as part of impressive efficiency advances over this period.

The product characteristics and the flow geography exert a strong influence on these consolidation decisions, and even more so on the degree to which even these consolidated routes are suitable for intermodal transport (Eng-Larsson and Kohn, 2012; Monios, 2015). An additional aspect indicated in Figure 9.2 is which actor controls the flows and hence manages the consolidation and movements. The broken line in the figure indicates the supplier managing the transport and the unbroken line denotes receiver control. This demonstrates the trend towards greater control by the receiver, for example large retailers who can lower costs by managing their own inbound transport, usually in partnership with a 3PL.

Product characteristics influence the number and location of DCs, such as the differences between non-food lines (e.g. electronics, homewares), ambient (e.g. tinned food), chilled, frozen and fresh food. Some product lines can be stored in a central NDC while others will go to regional and local DCs and some direct to stores. Those that are most suitable for intermodal transport are ideally not perishable and of lower value as there is danger of pilferage on overnight freight trains. Temperature-controlled goods can be transported with refrigerated rail containers but these are more costly hence rarer as the rail operator may not choose to invest in a fleet of such containers that are less flexible for other uses. They are also slightly wider which can be a challenge where loading gauge restrictions are present, for example in the United Kingdom. Finally, they are not compartmentalised, whereas road trailers can have more than one temperature zone for carrying a range of goods with different requirements. Fresh products such as milk and bread are not suitable for intermodal transport as they are delivered direct from suppliers to supermarkets rather than going via a DC. Flow volume refers not only to total volume but also to order frequency and size. Some goods (particularly fresh food) require regular replenishment while others require infrequent orders. The volumes most suitable for intermodal transport are large volumes with regular delivery requirements and ideally non-time sensitive. Predictable regular flows are ideal due to the longer lead time and infrequent schedule for intermodal services.

Flow geography is the interconnection between fixed sources of products (e.g. fruit from a particular region or the location of a factory producing certain products) and the DCs and/or stores. That is where PCCs can be useful, for consolidating supplier deliveries into one location from where the primary distribution to a DC can take place. If using intermodal transport, it is important to be able to consolidate flows at either end via some kind of hub-and-spoke system whereby the main haul is undertaken by rail. Consolidators or integrators such as 3PLs or other intermodal service providers are essential in creating a full train from a number of different customer shipments, regardless of whether they book the entire train from the rail operator and organise the slots themselves or merely book space on an existing service (see Chapter 8 for more discussion of intermodal transport business models).

In order to deal with the foregoing issues, it is therefore no surprise that a number of supply chain decisions need to be made for a firm to switch to intermodal transport, such as increased inventory, extended delivery windows and improvements in planning and ordering due to less flexible departure times (Eng-Larsson and Kohn, 2012; Monios, 2015). Increased inventory is needed because of less frequent deliveries. Given the increased trend towards 'just in time' in recent years, DCs and stores hold less inventory than in the past and rely on regular deliveries. Floor space can be much more productive by not keeping large inventories as was the norm in previous decades. Cross-docking is also preferred where possible as it saves time to keep loads in movement rather than storing in racks that then have to be picked later. Cross-docking is a practice whereby goods are unloaded into the warehouse and are loaded onto new trucks either directly or a short time later. For instance, 10 trucks may arrive at a DC from individual suppliers each fully loaded with a different product, but each store may require two pallets of each product. The pallets are unloaded from the inbound trucks into the DC, and two pallets of each product are then loaded onto each truck on the opposite loading bays destined for individual stores. Sometimes, this does not take place immediately and the combined loads will be prepared in front of each loading bay waiting for the trucks to arrive.

Intermodal transport is not only less frequent (e.g. perhaps only one service per day), but the deliveries are larger. There is a significant difference between a regular feed of single trucks to a DC and an entire train replacing the equivalent of 40 trucks all arriving at one time. Buffer stock may also be desirable in case of delays and cancelled services. Hence, using intermodal transport may require an increase in inventory leading to an increase in cost, which is unlikely to be offset by a corresponding decrease in transport cost from changing mode.

Delivery windows determine the availability of DC staff to unload the incoming vehicles, and are also linked with other aspects such as the timing for outbound trucks to pick up the re-sorted and picked goods from the inbound trucks. In order to minimise staff costs as well as ensure that inventory does not pile up, DC operations often provide drivers with specific delivery windows during which they must arrive. If they arrive outside this window then there is a high possibility that the load will be refused, which increases costs for all concerned as the load will need to go back to the supplier and redelivered at a later agreed time. Switching to intermodal transport often involves extending the delivery window in order to deal with the increased size of deliveries, which are necessitated by the lower frequency of intermodal services. Extending the delivery window will require more staff and more space. On the other hand, the total delivery window may be shorter because several trucks' worth are dealt with at one time rather than the time that would have been taken to handle one truck after another throughout the day or week. Larger drops will also require several loading bays which may not be present in smaller DCs. In some cases, the container or trailer can be left at the warehouse for emptying later, but this will incur its own costs of marshalling trailers or containers on dollies or needing to have their own handling equipment to pick containers off the ground to loading bay height.

Linking all of the above is the need for system control, usually in-house by the shipper or in combination with a 3PL or similar who can coordinate the transport chain, amalgamate flows and minimise disruption. As noted earlier, the roles of vertical and horizontal integration have been instrumental in linking shippers, 3PLs (or other intermodal consolidators) and rail operators in order to establish traffic on these routes. The relations between distribution (linking retailers and intermodal transport provider) and intermodal transport (linking intermodal transport provider and rail operators) are depicted in the conceptual framework

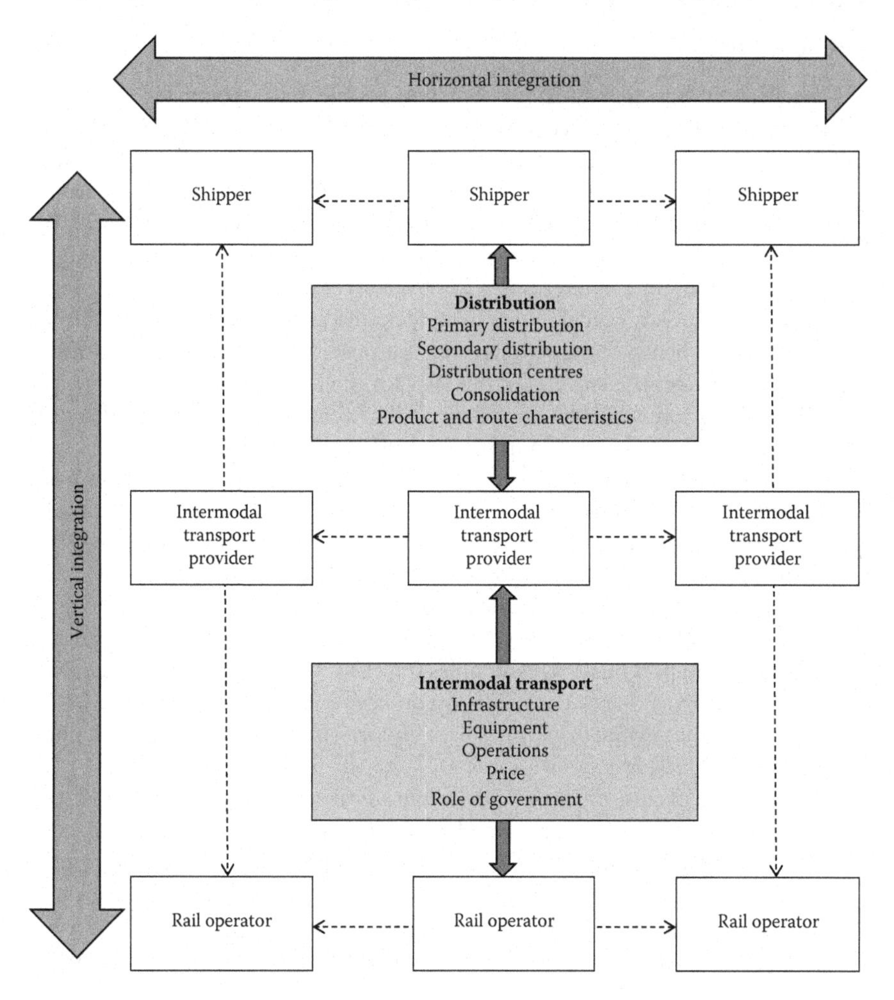

Figure 9.3 Conceptual framework linking distribution characteristics, intermodal transport provision and the role of integration. (Adapted from Monios, J., *Transportation Planning and Technology*, 38(3), 1–28, 2015.)

developed by Monios (2015; Figure 9.3). The roles of vertical and horizontal integration have been foregrounded as key elements.

The importance of 3PLs acting as consolidators is key, but in the diagram the term *intermodal transport provider* has been used as this company may in fact be a different kind of operator (see Chapter 8 for full discussion). The important issue is that it is an intermediary company that coordinates the shippers' needs with the physical rail operations performed by the rail operator. Indeed, in some cases it will be the rail operators themselves dealing with the shipper.

CASE STUDY OF UK RETAILERS*

UK retail sector

UK retailers employ approximately 3 million people and account for almost 6% of UK gross domestic product (GDP) (Forum for the Future, 2007; Jones et al., 2008). Nearly 83% of the retail market of the grocery trade in the United Kingdom is controlled by five supermarket retailers: Tesco (31%), Asda (17%), Sainsbury (16%), Morrison (12%) and the Co-operative (7%) (Scottish Government, 2010). While the industry deals with external pressures such as market saturation, competition and demographic shifts (Kumar, 2008), market power has also been concentrated among a few large retailers due to mergers and acquisitions (Burt and Sparks, 2003). DCs are being optimised and new purpose-built facilities are appearing. Figure 9.4 illustrates the location of the DCs for the five major grocery retailers in the United Kingdom (PCCs are not shown). The centralisation in the Midlands is clear, as is the lack of coverage in north England, north Scotland and Wales.

This case study will look primarily at the Anglo-Scottish movements from DCs in the Midlands to Scotland, as representative of both successful intermodal transport and trends towards greater centralisation. McKinnon (2009: p. S295) found that 'since 2004, roughly 60% of the demand for large DCs has come from retailers'. Large firms are reducing the number of their DCs while increasing the size and efficiency of those that remain. Fewer, larger DCs means greater centralisation and potentially greater miles travelled, but also greater potential for intermodal transport due to consolidation on key routes. Food and grocery companies currently contribute one in four of all lorry miles travelled in the United Kingdom (IGD, 2012).

Greater use of information and communications technology (ICT) has allowed more accurate forecasting and more responsive ordering (thus a move from push to pull); filling these orders without incurring increased transport costs then requires a more tightly optimised spatial distribution of facilities, as well as greater integration between primary and secondary networks. Thus, some retailers work with hauliers to optimise their distribution (e.g. reducing empty

* This section draws on a larger case study described in Monios (2015). The empirical data was correct in 2012.

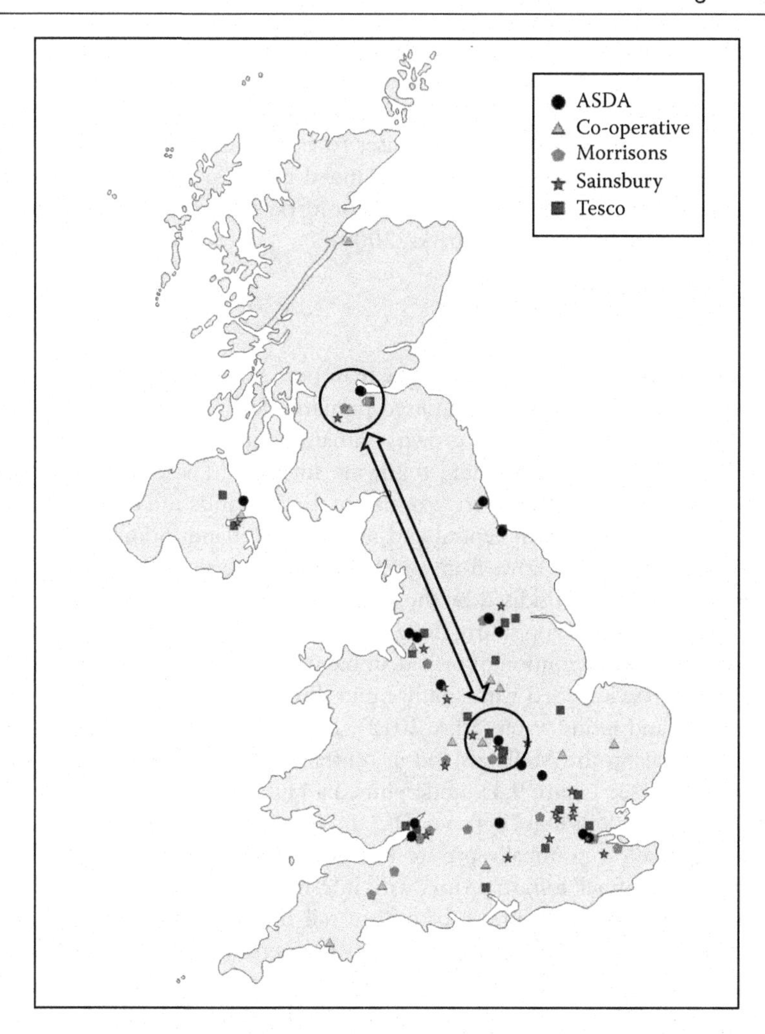

Figure 9.4 United Kingdom distribution centres of the top five supermarket retailers with Anglo-Scottish rail corridor marked 2012. (Based on Monios, J., *Transportation Planning and Technology*, 38(3), 1–28, 2015.)

running or reducing inventory holding requirements) or work with suppliers to optimise product flows (e.g. forecasting, planning and ordering). The result of these spatial and operational evolutions has been the increasing integration of operations, ranging from increasingly efficient use of backhauling to the implementation of factory gate pricing (FGP), both of which give the retailer greater control over primary distribution, thus strengthening its negotiating position (Mason et al., 2007; Potter et al., 2007; Burt and Sparks, 2003; Towill, 2005).

This period of industry evolution also saw increased use of 3PLs to handle the growing and increasingly complex transport requirements resulting from these developments, as well as more frequent, smaller deliveries from suppliers to reduce inventories, which also encouraged suppliers to make use of PCCs (Smith and Sparks, 2009; Fernie and McKinnon, 2003). Distribution facilities continued to evolve, from single product warehouses to composite environments housing ambient, chilled and fresh produce, all scanned in and out using barcodes that were integrated within the IT system used for forecasting, planning and ordering (Fernie et al., 2010; Smith and Sparks, 2009).

Use of rail

Domestic intermodal traffic took longer than port flows to establish, remaining marginal in the earlier years and utilised primarily for industrial products. Over the last decade, this market has grown, primarily due to retail flows, with Asda first using rail in 2003 and Tesco following in 2006. These flows are on the Anglo-Scottish corridor (between terminals in the Midlands and central Scotland) and intra-Scottish (between central and northern Scotland, primarily representing continuations of the flows from the Midlands services). These developments were to some extent subsidised by the successful use of government funding for intermodal terminals (Woodburn, 2007). These flows have been primarily northbound secondary distribution of picked ambient grocery loads from retail DCs in the Midlands, back loaded with southbound flows from Scottish suppliers, such as soft drinks and spring water (FTA, 2012). The concentration of DCs and intermodal terminals in the Midlands and in central Scotland, with suitable distance between them (see Figure 9.4), underpinned a high-density Anglo-Scottish corridor with a short 'last mile' between DC and intermodal terminal at either end.

Network Rail, a nominally private but government-owned company, owns and operates the track infrastructure, with intermodal terminals owned or leased by private operators. A number of private rail operators compete to run services. There are four primary rail freight operators in the United Kingdom: DB Cargo UK (formerly EWS then DB Schenker then rebranded as DB Cargo UK in 2016), Freightliner, Direct Rail Services (DRS) and GB Railfreight. The other main players are third-party logistics service providers that charter trains from these operators, including John G Russell, WH Malcolm and Eddie Stobart. There has been a significant growth in 3PLs because many customers prefer integrated door-to-door solutions.

Table 9.1 lists all current intermodal rail services on the Anglo-Scottish route (not only those used by retailers), divided into two categories: ex port (direct service between a port and a Scottish terminal) and domestic (between inland terminals in England and Scotland). Intra-England and intra-Scotland services are not shown. The table shows that the majority of domestic container rail traffic between Scotland and England uses DIRFT Daventry; currently doing around 175,000 lifts per year, it is the busiest inland intermodal terminal in the United Kingdom.

These intermodal services are all shared user. The ex port services are majority booked by shipping lines as carrier haulage is dominant in the United Kingdom

Table 9.1 List of current intermodal rail services running on the Anglo-Scottish route 2012

Type	*Service*	*Traction*	*Management*	*Frequency Per Week*
Port	Felixstowe-Coatbridge	Freightliner	Freightliner	5
	Southampton-Coatbridge	Freightliner	Freightliner	5
	Tilbury-Coatbridge	Freightliner	Freightliner	5
	Liverpool-Coatbridge	Freightliner	Freightliner	5
Domestic	Tilbury-Barking-Daventry-Coatbridge	DRS	JG Russell	2 daily × 5/6
	Daventry-Mossend (DB Schenker)	DB Schenker	Stobart	6
	Daventry-Mossend (PD Stirling)	DRS	WH Malcolm	5
	Daventry-Grangemouth	DRS	WH Malcolm	6/7
	Hams Hall-Mossend	DB Schenker	DB Schenker	5

for port flows, but smaller users can also book space on these trains directly with Freightliner or through a 3PL or freight forwarder. The other flows are managed by 3PLs serving a variety of customers. The largest sector utilising these trains is the retail sector. In this research, the focus is primarily on grocery retailers rather than other retail sectors such as fashion, and a wholesaler has also been included as a contrast.

Woodburn (2003: p. 245) noted that 'it is notoriously difficult to identify specific rail freight users and volumes from public sources, particularly in the non-bulk sectors'. For this research, a list of all retailer use of intermodal transport was compiled by combining the interview data and a report by the UK Freight Transport Association (FTA, 2012). The results are presented in Table 9.2.

Tesco is the only retailer large enough to move significant flows by rail, with four dedicated services, matching the secondary distribution of picked loads with inbound primary flows, filled out with other materials such as packed-down cages and recycling. Tesco transports 32 45 ft loads daily northbound on the Anglo-Scottish corridor, while their new service to Wales takes 34 45 ft boxes, and their service to the north of Scotland and the one from Tilbury take 22 containers each. Asda (not interviewed for this study) is the only company that gets close, with 20 loads on the Anglo-Scottish route and 10 going to Aberdeen. Tesco is about to start moving up to 20 loads daily on the Aberdeen route, as well as planning some more potential services in collaboration with DRS, only one of which is likely to be a dedicated service. With the additional Tesco volume, the Aberdeen service is now fully utilised and is about to extend to a seven-day operation. At the time of research, Tesco was planning some more potential services in collaboration with DRS, only one of which was expected to be a dedicated service.

Table 9.2 Use of intermodal transport by large retailers in the United Kingdom 2012

Retailer	*Route*	*Traction*	*Management*
Tesco	Anglo-Scottish	DB Schenker	Stobart
Tesco	Scotland to north	DRS	Stobart
Tesco	Daventry-Tilbury	DRS	Stobart
Tesco	Daventry-Magor	DRS	Stobart
Sainsbury	Anglo-Scottish	DRS	JG Russell
Morrison	Anglo-Scottish	DRS	JG Russell
Costco	Anglo-Scottish	DRS	JG Russell
Waitrose	Anglo-Scottish	DRS	WH Malcolm
M&S (DHL)	Anglo-Scottish	DRS	WH Malcolm
Co-operative	Anglo-Scottish	DRS	WH Malcolm
Asda	Anglo-Scottish	DRS	WH Malcolm
Asda	Scotland to north	DRS	DRS

Wholesaler Costco is the only other significant user of rail transport, sending 10–15 containers daily on the JG Russell service to Scotland. They used to send the Aberdeen deliveries on this train (just to Coatbridge then by lorry to Aberdeen), but because of the timings it was found to be quicker to use road. The train arrives early enough to suit the central belt stores but there would not be enough time to drive the deliveries up to the Aberdeen stores.

Other users only contribute very small numbers of containers to the shared user Anglo-Scottish services. Sainsbury has been using rail on some primary hauls to bring the products of Scottish suppliers to their Midlands DCs, using the shared JG Russell service (although management of this flow has recently returned to the supplier). Morrisons use the JG Russell service in the opposite direction to move loads of picked pallets from Northampton to Bellshill. In the past, they have trialled services between Trafford Park and Glasgow, and Coatbridge to Inverness. Waitrose uses the WH Malcolm Anglo-Scottish service, as does DHL for Marks and Spencer (M&S). M&S is building its own rail-connected DC at Castle Donington. The Co-operative is currently running a trial on the WH Malcolm Anglo-Scottish service, taking two containers per night, five nights/week from the Midlands to their Scottish DC at Newhouse.

The interviewees in this study claimed that asset utilisation is more important than break-even distance, even if made up by a number of short distance services. One interviewee said to 'beware of management accountants' because they look at the individual costs of running a train without considering factors such as utilisation and cross-subsidy across their service portfolio. Most freight trains run at night due to path restrictions during the day, with the result that a locomotive and wagon set may sit idle all day. Daytime running is generally possible in Scotland because the lines are not as busy but this is difficult in England. The view of rail operators in interviews conducted for this research is that if you can keep a train running most of the day then it will make money, so if a train is just sitting idle in a siding then any service, no matter how short, is worth running.

The handling charges necessitated by changing modes have always been a barrier to the greater use of intermodal transport, and the lack of visibility of the true cost of rail movements was noted in the literature review. One retailer suggested that the quote they are given is simply based on being 'slightly cheaper than road' rather than being based on the actual costs of providing the service; they would like greater visibility of the cost to the provider of the entire rail service, including the trunk haul. This is similar to the greater control over primary distribution sought through the use of FGP. It is a way of removing the need for the retailer to pay a profit margin on top of the base cost of the transport service. Retailers have been able to make intermodal transport more affordable by bargaining the handling price down, but rail operators feel that they cannot go any further or they will not be able to provide the service.

Retail movements to stores are generally done in cages, and greater economies can be achieved by transporting these in double-deck lorries, which are almost unique to the United Kingdom (McKinnon, 2010). A standard lorry takes 45 retail cages, as does a 45 ft rail container, whereas a double-deck lorry can take 72 cages. As confirmed by a retailer in an interview: 'because we run double-deck road trailers, it is difficult for the rail operators to compete on price'. Double-deck lorries currently form about 20% of the Tesco fleet. That might eventually get up to around 40%–50%, but, according to the interviewee, it will never be 100% due to operational reasons.

Obtaining flows in both directions is often the key issue in making intermodal transport economically competitive with road. By integrating their primary and secondary distribution, Tesco has been able to match supplier deliveries inbound to their Daventry NDC with outbound distribution to regional distribution centres (RDCs). For example, they sell space on their dedicated trains to their suppliers, thus inserting themselves in a chain of vertical cooperation that draws the rail operator, 3PL and supplier together. JG Russell matches flows on rail by sending French wine to Daventry, then the Costco loads to Scotland, then returning from Scotland to the continent with whisky.

More backhauls from Scotland to England are needed to support the Anglo-Scottish services. The feeling from the interviewees is that the loads are there; 'it is just a matter of making it work', sometimes just convincing a company that has not used rail before to give it a try. 3PLs feel that there are many companies with a few containers a day that could use rail, or that may require consolidation of less-than-container loads (LCL) before sending them south by rail. Therefore, consolidation could be a key issue to promote further use of intermodal transport and integrate road and rail more seamlessly.

Road operations also need to be understood in order to contribute to supporting the growth of intermodal transport. Road haulage is built into supply chains because of its inherent flexibility, for instance the ability to stagger deliveries. If

30 containers arrived together it could be difficult to handle. 'Staggered delivery is easier to manage', one retailer said.

Distribution requirements versus intermodal transport characteristics

The Anglo-Scottish corridor in the United Kingdom is an excellent example of the warehouse and transport cost trade-off discussed earlier in the chapter. Not all retailers have the resources or the desire to manage primary distribution, as there are pros and cons to managing it in-house, sub-contracting to one or more firms or leaving it to suppliers. For example, Sainsbury manages about 90% of their inbound fresh produce, 60% of chilled and 10% of ambient/grocery, whereas Tesco has a larger focus on primary distribution, with 60% of ambient/grocery and 70% of fresh produce moving through their primary network. Tesco's high level of control of primary ambient flows enables them to put this supplier traffic on rail, providing backhaul flows south to the Midlands to balance the northbound secondary movements.

These decisions are different for different companies, thus a wholesaler such as Costco has a simpler model. They only have about 3200 stock-keeping units (SKUs), so this is very different from a supermarket retailer, as it allows Costco to maintain a far simpler operation. All primary flows are delivered as full container loads (FCL) and managed by the suppliers; moreover, all value-added work is pushed upstream as supply chain management is not the core competency of Costco.

Tesco, Sainsbury and the Co-operative all run their own trucks for secondary distribution but will sub-contract occasionally where required, as well as some of the retail lorry fleet operated by third parties on an open-book basis. A large supermarket would have on average about five to six trucks delivering per day (obviously this depends on a number of factors such as the use of double-deck lorries).

Whether secondary flows are suitable for rail will depend to a large degree on the distribution strategy of the retailer; for example, which product lines are stored at the RDC and which require trunking from the NDC. For example, when Tesco moves containers by rail from Daventry to Livingston, each container is designated for a specific store, with the relevant cages from Daventry inside. At Livingston, they add additional cages to the container, then send it by truck to the store. This is done in the trunking station which is all cross-docked. The Stobart Tesco train to Inverness also takes boxes for specific stores, but rather than being a DC to DC move, these boxes are trucked direct from the terminal to stores by JG Russell.

Opened in 2007, Tesco's large one million square foot DC at Livingston is the only Tesco DC that has fresh, grocery, frozen, trunking and recycling all within the same facility. It sends around 4.5 million cases weekly to about 250 stores across Scotland, north England and Northern Ireland. There are 7500 SKUs in the grocery part of Livingston DC alone. Tesco monitors which lines should be picked at Daventry and trunked to Livingston and which should be stored there. It changes as different lines rise and fall in sales; however, all fresh food in Scotland moves through Livingston. On an average day, the Livingston DC has around 900 trucks coming in and out, but this is an unusually large site.

Lead time is crucially important for all movements between DCs and stores. According to interviewees, an ideal scenario would involve overnight picking and morning departure from the DC to reach the store by mid-afternoon, but this cannot always be done because of passenger trains on the line during the day. Unless this can be resolved, intermodal growth will be constrained by operating mostly at night, which requires stores to order from DCs in the morning so that the load can be picked in the afternoon, loaded at the DC at say 16:00 to catch a 20:00 departure on the train, which will then arrive at its destination in the early morning (say 04:00–05:00) for trucking to the store.

Integration and collaboration

Both vertical and horizontal integration can be observed in the industry; however, it is the former that is having the greatest impact. Most noticeable is the relationship between the retailer Tesco, the logistics provider Stobart and the rail operator DRS. Working closely together has allowed all parties to develop knowledge of requirements and adjust operations to suit as they plan new services and solve operational issues as they arise.

Different models of vertical integration between levels of rail operations (from terminal operation, traction provision, train management and road haulage) are discussed in Chapter 8. Terminals can be run by rail operators (e.g. Freightliner or DB Schenker), 3PLs (e.g. WH Malcolm) or other companies (e.g. ABP at Hams Hall), or even be private sidings for which the operation is sub-contracted (e.g. Stobart operating the Tesco siding at DIRFT). Likewise, the customer side of trains is normally managed by a 3PL rather than the rail operator (e.g. JG Russell, WH Malcolm and Stobart operating trains with traction provided by DRS or DB Schenker), but for other trains the management is also done by the rail operator (e.g. Freightliner or DB Schenker).

Retailers do not currently share space within containers, but their containers travel together on multi-user 3PL trains as noted earlier. 3PLs also share space on their services, usually on an ad hoc basis, but as port services are mostly run by Freightliner who specialises in these flows, 3PLs will buy space on those trains (e.g. Stobart bringing boxes from Tilbury to their hub at Widnes). 3PLs can collaborate in other ways, for example, Stobart runs the Tesco train from Mossend to Inverness, where it terminates at the JG Russell terminal, from which point JG Russell distributes the containers to stores by road. However, while 3PLs will share space on each other's services when needed, they do not actually run any regular services together.

Currently, Tesco is the only retailer large enough to fill a complete train. However, the decision is whether to operate a dedicated service, in which case the retailer pays for the whole train and therefore must take responsibility for filling any empty wagons or suffer a financial penalty. Scheduled services may be used by any shippers, but having a dedicated train grants more control over the timings and operation of the service. Establishing a dedicated Tesco train rather than just buying space on a third-party service gives them more control and enables them to plan the primary and secondary distribution as part of a unified system.

The retail interviews revealed that in an ideal scenario they would all prefer to have their own rail-connected sheds rather than using a shared terminal to load a multi-user train. This practice would require a full trainload per user, reducing opportunities for collaboration. The new DIRFT 3 appears to be planned around this model of more rail-connected sheds. Similarly, the new DC being built by M&S at Castle Donington is rail-connected; however, without the retailer being able to provide enough volume for regular services, this development will make asset utilisation more difficult for rail operators. It can only work if an operator (or someone else) can provide more rail flows to this terminal to get better utilisation of the rolling stock.

CASE STUDY OF SHIPPER-FORWARDER INTEGRATION BY JULA AND SCHENKER*

Company background

Jula operates in the DIY segment and focuses on offering professionals an attractive range at low prices. This is possible through large purchases directly from manufacturers all over the world, without intermediaries. The product range has expanded over the years to include tools, equipment, work clothing, garden products, paints and household items. As of 2014, the company had 73 department stores in three countries (Sweden 41, Norway 21, Poland 11) and 2400 employees. The 2013 company turnover was €0.5 billion with profits reaching €57 million. The company has a strong equity ratio of 48% (2013). All flows are consolidated at its 150,000 m^2 central warehouse and DC in Skara. The majority of the incoming goods to the central warehouse consist of imported containers, mainly from Asia. Schenker Air and Ocean in Sweden hold the Jula key account and coordinate incoming container flows.

Initiative

Jula and Schenker Air and Ocean had collaborated closely for more than a decade before the discussions regarding a joint intermodal transport service started. The first idea for an intermodal transport service came from the municipality

* This section draws on a larger case study described in Monios and Bergqvist (2015).

of Falköping, which did a pre-study to analyse the possibilities of a rail shuttle between the Port of Gothenburg and the intermodal terminal at Falköping. The study proved that there was environmental and cost-saving potential as well as service quality improvement possibilities given that the container flow could be managed much more efficiently by using the terminal in Falköping as a buffer for full containers as well as an empty container depot, meaning that containers could be more easily distributed from the terminal in Falköping to exporting companies in the region. At the time, empty containers were often shipped back to the Port of Gothenburg and then repositioned to exporting companies. Jula was experiencing rising storage costs for full containers at the Port of Gothenburg and actually repositioned containers to a nearby container depot in Gothenburg. In order to achieve the identified potential, however, a substantial share of the container flows in the region had to be coordinated and consolidated on the intermodal rail service.

The study outcome was presented to Jula management in 2011, who responded positively but wanted Schenker to be part of the intermodal transport solution. Another issue for Jula was that they always enjoyed cheap road haulage because they had the largest container flows in the region and their dominant import flows were attractive to road hauliers when trying to match import and export container flows. The study revealed that the intermodal transport solution could be competitive with around 10,000 TEU (20 ft. equivalent units) per year (cf. Ye et al., 2014), which was a little less than Jula transported during 2011, even considering the company's steady annual growth of about 10%–15%.

It was not until 2012 that Jula's volumes had increased to the point where they could potentially make up the critical mass for a profitable and stable intermodal transport service. Schenker and Jula established a joint project team to realise the idea in January 2013. After about 1 year of preparations and investigations, the intermodal transport service was launched, with the first train departing from the container terminal of the Port of Gothenburg (Skandiahamnen) for the inland terminal at Falköping on 4 September 2013. The service started with a 'half train' of 11 wagons, with a capacity of 44 TEU in each direction. As of October 2014, the train capacity was increased to 17 wagons, carrying 68 TEU. As of 2017, the train operates at maximum length, that is with 21 wagons carrying 84 TEU in each direction. Since the start, the intermodal transport service has operated five times per week.

Stakeholders and contracts

Although Jula's volumes increased and critical volume was achieved around 2012, there was a long journey ahead to coordinate all stakeholders in order to develop the necessary intermodal terminal facilities and to sign contracts in a synchronised manner and with long enough contract periods to make stakeholders willing to invest. Figure 9.5 illustrates the complexity in terms of the number and type of agreements and the fact that they had to be coordinated and

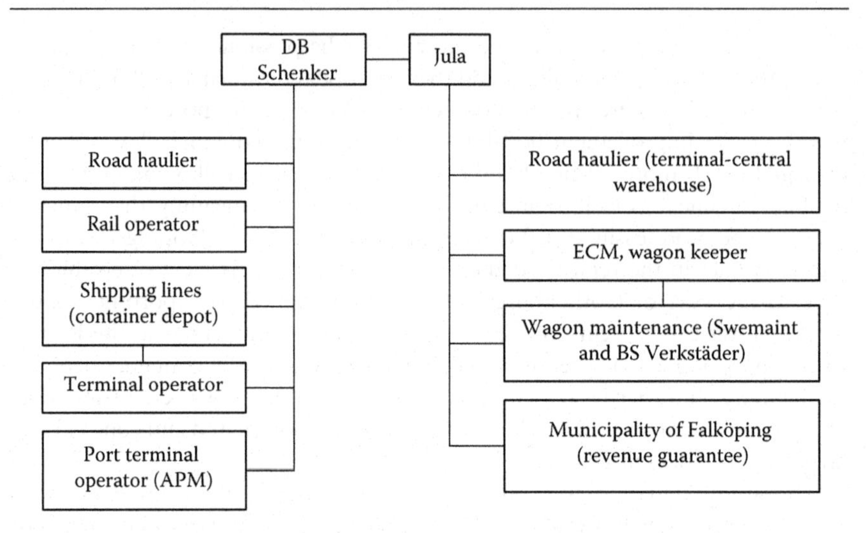

Figure 9.5 The structure of agreements. (From Monios, J. and Bergqvist, R., *Supply Chain Management: An International Journal*, 20(5), 534–548, 2015.)

synchronised. Furthermore, the agreements were followed by a long process of trust building in order for stakeholders to establish enough confidence and willingness to invest – this refers to both private and public actors.

The central agreement is the one between Jula and Schenker with a focus on defining how risks, investment and benefits are distributed. They operate an open-book agreement with a very high level of transparency and both actors are involved in discussions covering aspects such as pricing, investments, service quality and tendering processes. Both Schenker and Jula have recognised the importance of signing long-term contracts in order to incentivise the terminal operator to invest in the required handling equipment and the municipality to invest in a new terminal adjacent to the old terminal. That is the reason why Schenker signed a 2-year contract with the rail operator and a 5-year contract with the terminal operator (the terminal operator was appointed by the municipality of Falköping through the process of public tendering).

The Swedish rail system has been deregulated since 1988 and there is substantial competition in the rail market; therefore, Schenker and Jula saw it as unnecessary to run the train themselves. In addition, they wanted to explore opportunities for creative suggestions that might be offered by existing operators. They ran a tender allowing the rail operators to suggest different solutions whereby the Gothenburg–Falköping rail shuttle could be combined with other rail solutions and destinations; hence, the timetable was not entirely fixed but an indication of favourable time windows was given.

To enable long-term investment in a new intermodal terminal by the municipality of Falköping, Jula signed a separate agreement guaranteeing revenues of

€250,000 for the intermodal terminal for a period of 5 years, starting 1 January 2014. Annual variable terminal rent fees (about €4 per handled container) are balanced against the guaranteed revenue if Jula were to end the deal within the 5 year period. This agreement has been crucial in order for the municipality to invest about €2.5 million in developing a new intermodal terminal. A critical concern was to develop the rail shuttle in such a way that Jula and Schenker remain flexible and independent so that the sub-contracted rail operator does not gain too much power; this is often the case because they generally own the wagons and control the timetable and the time window (slot) in the container terminal at the seaport. In this case, Schenker has signed an agreement with the port container terminal operator APM Terminals and Jula has invested in container wagons (type Lags071 and SGNSS). Becoming a wagon owner means that Jula had to contract an entity in charge of maintenance (ECM) and a maintenance provider (Swemaint and the local service provider BS Verkstäder). The ECM provides evidence of responsibility and traceability of the maintenance undertaken on freight wagons in accordance with EU regulation EU/445/2011.

The timetable for the train cannot be controlled exactly since a rail traffic certificate is needed, which neither Schenker Air & Ocean nor Jula possess. Overall, the structure of agreements is rather complex, but by engaging with all interfaces a service set-up based on transparency and long-term commitment has been achieved. It could be argued that such a result is necessary in order to develop cost-competitiveness on an intermodal service over such a short distance.

Results of initiative

Schenker takes the responsibility for three main functions: management bookings, accounting and monitoring. Besides the operating functions, Schenker also has the responsibility for marketing and sales of the intermodal service to attract other shippers besides Jula. Schenker and Jula continuously discuss market issues since the aim is for Schenker and Jula to attract complementary flows, meaning customers with export flows and largely with the same shipping lines as Jula. This enables the effective repositioning of containers and high utilisation rates on the intermodal service. This also means that Schenker does not merely sell capacity on the intermodal service but takes full responsibility for the customers' export and import flows in order to be able to fully coordinate the usage of the service. Other customers that have since joined the intermodal transport service include companies such as Parker Hannifin, Swedish Match, A Lot of Decoration and Gyllensvaan (supplier of 'Billy' bookshelves to IKEA).

For the purpose of effective repositioning of empty containers, Schenker and the inland terminal operator signed agreements with shipping lines in order for them to set up an empty container depot in Falköping, which became more time-consuming and challenging than expected according to the representatives of Schenker. In addition, Jula has developed their customs clearance process so that the containers/goods do not need to be cleared until they reach the Jula warehouse in Skara.

Overall, the following benefits have been achieved as compared with the previous road-based transport service:

- Cost-efficiency.
- Traffic safety (less heavy transport on road).
- Environmental performance (about 80% less emissions of CO_2 vs. road transport).
- No waiting times at the Port of Gothenburg.
- No port demurrage and no road toll fee.
- Imported container stock now closer to Jula's DC/warehouse which creates more even cargo flow into the DC.
- Long-term agreements.
- Jula is seen as a good benchmark in the Skaraborg region. The new set-up creates opportunities for the entire region and the development of intermodal solutions.
- More efficient road haulage through exemption for long vehicles (32 m = 2 × 40 ft).

The final point relates to the project initiated by Jula to develop the possibilities of road haulage of two 40 ft containers simultaneously. This has had a significant impact on the cost-efficiency of the intermodal transport solution for Jula as about 70% of their containers are 40 ft containers and about 30% are 20 ft containers (cf. Bergqvist and Behrends, 2011 and see discussion in Chapter 5). The current road restrictions only allow for the simultaneous haulage of one 40 ft and one 20 ft container.

Jula started the process of applying for an time-limited exemption of 3 years to the current road restrictions for transport between the intermodal terminal in Falköping and the central warehouse in Skara in 2012, receiving final approval from the Swedish Transport Agency on 1 December 2014. One of the main arguments for the exemption is that it contributes to the efficiency of the intermodal transport solution and thus modal shift from road to rail. The road haulage project is one of the reasons why Jula chose to sign their own local road haulage agreement. Another factor was the need for a long-term contract in order to persuade the local road haulier to invest in a dozen chassis in order to handle the Jula container flows between the intermodal terminal and Jula's central warehouse.

Future developments, goals and challenges

Currently, Schenker and the terminal operator are focused on developing more agreements with shipping lines in order to increase usage of the container depot at the intermodal terminal at Falköping. Schenker also focuses on attracting more shippers to the intermodal transport service. This process is time-consuming since shippers are often locked in to existing 1–2 year agreements, but more customers are added continuously. The goal is to reach about 25,000–30,000 TEU annually (fully loaded containers in total for both directions) within 2–3 years; currently the service handles about 15,000 TEU annually (excluding empty containers).

Another aspect that will benefit the intermodal transport service is the current development of the container terminal in the Port of Gothenburg which will allow longer trains (up to 750 m) and generate many more time slots for train arrivals thus increasing overall capacity in the network.

The partners are also planning to add additional routes to make better use of the assets (locomotive and wagons). Possible new routes that have been identified relate to not only incoming flows of input material such as wooden plates to the region but also outgoing flows from the region, for example flows to the north of Sweden, Norway and Finland. This will, however, require a new agreement with the traction provider. This new initiative means that Jula and Schenker will gain better utilisation of their wagons and increased profit, enabling Jula to cross-subsidise its own transport costs even further with revenue earned not just on the Gothenburg–Falköping service but on additional routes as well. The sub-contracted rail operator will benefit from an additional contract but will not gain as much as it would were it to operate the new route itself in its own name. Thus, the introduction of a vertically integrated joint venture model affects the competitive market place of third-party rail operators competing for traffic. On the other hand, the efficiencies gained from vertical integration (including in this case the terminal infrastructure as Jula's long-term contract with the terminal enables efficient management and investment in the infrastructure) raise questions about the EU directive to separate infrastructure ownership from rail operations.

One important conclusion is that many stakeholders in this case share the need for a continuous improvement process that requires all stakeholders to remain committed to developing the service, value-added activities and infrastructure. The elements of entrepreneurship and trust are evident and the cooperative business model is crucial for the construction and maintenance of a sustainable win-win context.

From the perspective of Schenker, they now consider extending the concept to other regions and destinations; however, this requires the same long-term commitment and perspective on mutually beneficial relationships with key stakeholders such as large shippers/customers and transport service providers. This is currently the main challenge since few shippers are used to or willing to engage in this type of cooperative business model. Schenker hopes that the best practice illustrated by the Jula case will help convince shippers and other stakeholders of the potential associated with this kind of business model, which indeed underlines the need to identify and classify its key features.

CONCLUSION

This chapter has demonstrated how switching from road haulage to intermodal transport is not simply a transport decision but requires an appreciation of the logistics set-up of the individual shipper. It is especially important for cases where intermodal transport is not cheaper than road due to the structure of their business, and changes must be made to inventory levels, DC locations and staffing times in order to produce a flow geography more suited to intermodal transport.

The first case showed that centralisation has facilitated trunk hauls between NDCs and DCs, thus being a key reason behind the success of retail intermodal logistics; however, different practices were observed. Horizontal integration is important to achieve full trains, and while the results showed that this is now happening between 3PLs, it is not happening with retailers. Public planners might consider whether multi-user platforms should be preferred in the planning system rather than more rail-connected sheds. Vertical integration is more common than horizontal, as was to be expected, as actors see that this is an important element in making intermodal transport work. By studying the role of 3PLs as well as retailers, this case demonstrated the value of taking a broader approach to the support of intermodal service development.

3PLS seem confident that intermodal use will increase; however, while retailers are positive about intermodal, actions show that it remains a minority interest. Indeed, it was unclear to what degree a company's interest in using rail is due to a shift in the sector or a purposeful management policy or whether it is just down to an individual in a company. Therefore, it is difficult to drive this through policy when it often comes down to individuals, meetings and discussions between 3PL or rail personnel and the potential client, built on individual relationships.

While the first case demonstrated the importance of 3PLs for linking shippers with rail operators, the second case revealed how a unique business model can be used to integrate a shipper and a 3PL to provide their own rail service by sub-contracting a traction provider. Although in its early stages of development, results indicate several advantages of this model, including risk sharing, knowledge development, long-term service stability and diversification of activities which all contribute to facilitating the shift of the large customer from road haulage to intermodal transport. Potential challenges mainly relate to contractual and relationship complexity. By the direct involvement of the shipper in the service design, the benefits achieved went beyond just cost reduction to include greater strategic control over the service and future development possibilities. The forwarder Schenker is considering expanding the service to another route serving different shippers with the use of Jula's wagons and the profits from this service will cross-subsidise the original route. Therefore, the two partners are acting like a new entity.

The overall conclusion from both cases is for policymakers and planners seeking to encourage the adoption of intermodal transport to understand that it is not simply funding of infrastructure that is needed but business restructuring. In many cases, this can be achieved without additional cost but in others some costs will be incurred. Nevertheless, it is essential that such costs are understood from the outset so that clear decisions can be made and unforeseen cost increases or service quality decreases do not occur.

REFERENCES

Abrahamsson, M., Brege, S. (1997). Structural changes in the supply chain. *International Journal of Logistics Management.* 8 (1), 35–44.

Ballou, R. H. (1987). *Basic Business Logistics.* Prentice Hall. Englewood Cliffs, NJ.

Bärthel, F., Woxenius, J. (2004). Developing intermodal transport for small flows over short distances. *Transportation Planning and Technology.* 27 (5): 403–424.

Bergqvist, R., Behrends, S. (2011). Assessing the effects of longer vehicles: The case of pre- and post-haulage in intermodal transport chains. *Transport Reviews.* 31 (5): 591–602.

Burt, S. L., Sparks, L. (2003). Power and competition in the UK retail grocery market. *British Journal of Management.* 14 (3): 237–254.

Christopher, M. (2011). *Logistics and Supply Chain Management.* 4th edn. Prentice Hall. Englewood Cliffs, NJ.

Coyle, J. J., Bardi, E. J., Langley, C. J. (2003). *Management of Business Logistics.* 7th edn. Southwestern Publishing. Mason, OH.

Eng-Larsson, F., Kohn, C. (2012). Modal shift for greener logistics: The shipper's perspective. *International Journal of Physical Distribution and Logistics Management.* 42 (1): 36–59.

Fernie, J., McKinnon, A. (1991). The impact of changes in retail distribution on a peripheral region: The case of Scotland. *International Journal of Retail and Distribution Management.* 19 (7): 25–32.

Fernie, J., McKinnon, A.C. (2003). The grocery supply chain in the UK: Improving efficiency in the logistics network. *International Review of Retail, Distribution and Consumer Research.* 13 (2): 161–174.

Fernie, J., Pfab, F., Merchant, C. (2000). Retail grocery logistics in the UK. *International Journal of Logistics Management.* 11 (2): 83–90.

Fernie, J., Sparks, L., McKinnon, A. C. (2010). Retail logistics in the UK: Past, present and future. *International Journal of Retail and Distribution Management.* 38 (11/12): 894–914.

Forum for the Future. (2007). *Retail Futures; Scenarios for the Future of UK Retail and Sustainable Development.* Forum for the Future. London.

FTA (2012). *On Track: Retailers Using Rail Freight to Make Cost and Carbon Savings.* FTA. London.

Hingley, M., Lindgreen, A., Grant, D. B., Kane, C. (2011). Using fourth-party logistics management to improve horizontal collaboration among grocery retailers. *Supply Chain Management: An International Journal.* 16 (5): 316–327.

IGD. (2009). UK Food & Grocery Retail Logistics Overview. Available at http://www.igd.com/Research/Supply-chain/UK-food--grocery-retail-logistics-overview/ (accessed 24 September 2012).

IGD. (2012). Over 200 million food miles removed from UK roads. Available at http://www.igd.com/About-us/Media/IGD-news-and-press-releases/Over-200-million-food-miles-removed-from-UK-roads/ (accessed 11 April 2012).
Jensen, A., Sorkina, E. (2013). Customer managed intermodal transport solutions: Why, how, for whom? *World Review of Intermodal Transportation Research.* 4 (1): 37–54.
Jones, P., Comfort, D., Hillier, D. (2008). UK retailing through the looking glass. *International Journal of Retail and Distribution Management.* 36 (7): 564–570.
Kumar, S. (2008). A study of the supermarket industry and its growing logistics capabilities. *International Journal of Retail and Distribution Management.* 36 (3): 192–211.
Lemoine, O. W., Skjoett-Larsen, T. (2004). Reconfigurations of supply chains and implications for transport: A Danish study. *International Journal of Physical Distribution and Logistics Management.* 34 (10): 793–810.
Mason, R., Lalwani, C., Boughton, R. (2007). Combining vertical and horizontal collaboration for transport optimisation. *Supply Chain Management: An International Journal.* 12 (3): 187–199.
McKinnon, A. (2009). The present and future land requirements of logistical activities. *Land Use Policy.* 26 (S1): S293–S301.
McKinnon, A. (2010). *Britain without Double-Deck Lorries.* Heriot-Watt University. Edinburgh, UK.
Monios, J. (2015). Integrating intermodal transport with logistics: A case study of the UK retail sector. *Transportation Planning and Technology.* 38 (3): 1–28.
Monios, J., Bergqvist, R. (2015). Using a 'virtual joint venture' to facilitate the adoption of intermodal transport. *Supply Chain Management: An International Journal.* 20 (5): 534–548.
Monios, J., Bergqvist, R. (2016). *Intermodal Freight Terminals: A Life Cycle Governance Framework.* Routledge. Abingdon, UK.
Potter, A., Mason, R., Lalwani, C. (2007). Analysis of factory gate pricing in the UK grocery supply chain. *International Journal of Retail and Distribution Management.* 35 (10): 821–834.
Runhaar, H., van der Heijden, R. (2005). Public policy intervention in freight transport costs: Effects on printed media logistics in the Netherlands. *Transport Policy.* 12 (1): 35–46.
Schmoltzi, C., Wallenburg, C. M. (2011). Horizontal cooperations between logistics service providers: Motives, structure, performance. *International Journal of Physical Distribution and Logistics Management.* 41 (6): 552–576.
Scottish Government. (2010). *Food and Drink in Scotland: Key Facts 2010.* Scottish Government. Edinburgh, UK.
Slack, B., Vogt, A. (2007). Challenges confronting new traction providers of rail freight in Germany. *Transport Policy.* 14 (5): 399–409.
Smith, D. L. G., Sparks, L. (2009). Tesco's supply chain management. In: Fernie, J. and Sparks, L. (Eds.), *Logistics and Retail Management*, 3rd edn. Kogan Page, London, pp. 143–171.
Towill, D. R. (2005). A perspective on UK supermarket pressures on the supply chain. *European Management Journal.* 23 (4): 426–438.
Van der Horst, M. R., De Langen, P. W. (2008). Coordination in hinterland transport-chains: A major challenge for the seaport community. *Maritime Economics and Logistics.* 10 (1–2): 108–129.

Van Schijndel, W. J., Dinwoodie, J. (2000). Congestion and multimodal transport: A survey of cargo transport operators in the Netherlands. *Transport Policy*. 7 (4): 231–241.

Woodburn, A. (2003). A logistical perspective on the potential for modal shift of freight from road to rail in Great Britain. *International Journal of Transport Management*. 1 (4): 237–245.

Woodburn, A. (2007). Evaluation of rail freight facilities grants funding in Britain. *Transport Reviews*. 27 (3): 311–326.

Ye, Y., Shen, J., Bergqvist, R. (2014). High capacity transport associated with pre- and post-haulage in intermodal road-rail transport. *Journal of Transportation Technologies*. 4 (3): 289–301.

Chapter 10

Legal aspects of multimodal transport

Abhinayan Basu Bal

INTRODUCTION

During the twentieth-century modernisation of transport infrastructure, technological advances and new management techniques changed the traditional patterns of carriage of goods around the world. The most important change happened with the advent of containers, which revolutionised the international distribution of goods through better handling operations, storage and transportation. Containers made transportation affordable and transformed the world economy. Secondly, logistics started to play an important role in the manufacture and transportation of goods, and also in containerisation, particularly in areas related to handling, positioning, storing, transporting and maintenance of containers, whether full of cargo or empty. Thirdly, just-in-time delivery that aims to reduce inventory cost and save space for the storage of raw materials and manufactured and processed goods became relevant for modern transport operations as time is as important as distance in the assessment of transportation costs. Fourthly, the development of port infrastructure to serve as regional gateways made transhipment an important aspect of multimodal transport using the hub-and-spoke model.

The international transport industry has embraced multimodalism and door-to-door transport based on the efficient use of all available modes of transportation by air, water and land. However, there exist problems, especially concerning the determination of liability for lost or damaged cargo in light of the varying liability schemes governing the respective modes of transport and the often difficult task of identifying the modal leg during which the loss, damage or delay occurred. This engenders lack of legal certainty and predictability for the parties involved in multimodal transport.

In this chapter, the evolution of multimodal transport law is briefly presented followed by a discussion on multimodal transport contracts, relevant transport documents and the role of the multimodal transport operator (MTO) including relevant definitions and terminology widely used in legal and technical contexts. This chapter also examines some of the characteristic legal problems relevant to multimodal transport and discusses network and uniform liability systems commonly used in various international conventions and non-binding instruments.

EVOLUTION OF MULTIMODAL TRANSPORT LAW

Historically, 'tackle-to-tackle' carriage has been common practice in maritime transport. It means that the goods are carried by the shipper to the port, loaded on board a vessel and discharged at the port of destination. Thus, the carrier's obligations cover only the time when the goods are carried on the vessel. Notably, the Hague Rules* and the Hague–Visby Rules† are based on tackle-to-tackle delivery. Under Article I (e) of the Hague/Hague–Visby Rules, 'carriage of goods' covers the period 'from the time when the goods are loaded on to the time they are discharged from the ship'. The drafters of the Hamburg Rules‡ extended the scope of application to 'port to port'. Pursuant to Article 1(6) of the Rules, the carrier undertakes against payment of freight to carry goods by sea 'from one port to another'. The Multimodal Convention§ introduced a 'door-to-door' delivery in Article 1(1) by covering the carriage 'from a place in one country' at which the goods are taken in charge by the MTO 'to a place designated for delivery situated in a different country'. This convention created the concept of the MTO and the general basis of liability was conceived along the lines of the Hamburg Rules.¶

The Multimodal Convention was not the first unsuccessful attempt to establish a widely acceptable international legal framework for multimodal transport. The International Institute for the Unification of Private Law (UNIDROIT) started work on the subject in the 1930s, which resulted in the 'draft Convention on the International Combined Transport of Goods' in 1963.** This was followed by the preparation and adoption by the Comité Maritime International (CMI) of the draft Tokyo Rules in 1969.†† The draft conventions prepared by UNIDROIT and CMI were combined into a single text in 1970, under the

* International Convention for the Unification of Certain Rules Relating to Bills of Lading, 25 August 1924, 120 L.N.T.S. 155 (Hague Rules).

† Protocol to Amend the International Convention for the Unification of Certain Rules Relating to Bills of Lading 1924, 23 February 1968, 1421 U.N.T.S. 121 (Hague–Visby Rules).

‡ United Nations Convention on the Carriage of Goods by Sea, 31 March 1978, 17 I.L.M. 608 (Hamburg Rules).

§ United Nations Convention on International Multimodal Transport of Goods, 24 May 1980, U.N. Doc. TD/MT/CONF/16 (1980) (Multimodal Convention). The Multimodal Convention requires 30 contracting parties for entry into force but has so far received only 11 ratifications. Any success at reaching the required number of ratifications in the near future seems highly unlikely.

¶ See Multimodal Convention, Article 16, para. 1, noting that the MTO was to be liable for loss resulting from loss or damage to the goods, as well as from delay in delivery, if the occurrence which caused the loss, damage or delay in delivery took place while the goods were in its charge, unless the MTO could prove that it, its servants or agents had taken all measures that could reasonably be required to avoid the occurrence and its consequences.

** UDP 1963, ET.XL.II.DOC.29.

†† 10Draft Convention on Combined Transport, 1969.

auspices of the Inland Transport Committee of the United Nations Economic Commission for Europe (UNECE), known as the 'Rome draft'. This draft was further modified by meetings of the UNECE and the Intergovernmental Maritime Consultative Organization (IMCO) (now the International Maritime Organization [IMO]) during 1970 and 1971 which came to be known as the 'TCM draft'.* The TCM draft, unfortunately, did not proceed beyond the drafting stage. Its provisions were, however, subsequently reflected in standard bills of lading such as the Baltic and International Maritime Conference's (BIMCO) COMBICONBILL† and in the 'Uniform Rules for a Combined Transport Document' of the International Chamber of Commerce (ICC).‡

With ever-increasing demand for multimodal contracts, the international freight forwarding community, through its Swiss-based umbrella non-governmental organisation (NGO), the International Federation of Freight Forwarders Associations (FIATA), working with the ICC and the United Nations Commission on Trade and Development (UNCTAD) filled the vacuum left by the absence of a door-to-door regime with contractual provisions such as the UNCTAD/ICC Rules for Multimodal Transport Documents 1992.§ These rules apply where they are incorporated into a contract of carriage by the parties and help them avoid a multiplicity of different regimes governing multimodal transport. The fundamental premise of these rules is the concept of 'network liability', in which the carrier undertakes responsibility for the entire period of transport but to the extent provided by the applicable international convention or national law governing the particular segment or mode of transport where the loss or damage occurred. These rules have been described as flexible and practical by the industry, which can be used by MTOs as the basis for a multimodal transport contract. At the same time, operators have the option to add their own clauses on matters such as optional stowage, liens and so on, that are not fully within the scope of the UNCTAD/ICC Rules.

MULTIMODAL TRANSPORT CONTRACT AND MULTIMODAL TRANSPORT OPERATOR

A contract for sale is the starting point of any international commercial transaction and it is closely interrelated with other contracts relating to carriage of

* Draft Convention on the International Combined Transport of Goods. TCM refers to the French acronym for *Transport Combiné de Marchandises.*

† These are standard forms of contract which have terms providing for combined transport. Another example is MULTIDOC 95 bill of lading. See Appendix.

‡ Uniform Rules for a Combined Transport Document, ICC publication No. 273, ICC, Paris, 1973. They were slightly revised in October 1975 (ICC publication No. 298) (ICC Rules 1975) to overcome practical difficulties of application concerning the combined transport operator's liability for delay. The ICC Rules were conceived as an essential measure to avoid a multiplicity of documents for combined transport operations.

§ UNCTAD/ICC Rules for Multimodal Transport Documents, ICC publication No. 481, ICC, Paris, 1992 (UNCTAD/ICC Rules 1992).

goods, insurance and intermediated trade financing (see Bridge, 1999: p. 2; Ramberg, 2011: p. 37). The Incoterms prepared by the ICC in 1936 are widely used by parties as the contract for sale as they contain standard shipping terms.* The Incoterms consist of three-letter acronyms describing the tasks, costs and risks involved in the delivery of goods from sellers to buyers. The standard terms range from EXW (ex works) imposing on the seller to make the goods available for collection by the buyer, all the way to DDP (delivered duty paid), where the seller undertakes the cost and risk of actually delivering the goods to the buyer at destination. In between these extremes lie the F and C terms, such as FOB (free on board) and CIF (cost, insurance and freight) sales, with risk passing at the load port but costs to destination borne by the buyer in the first instance and the seller in the second. The terms EXW, FCA (free carrier), CPT (carriage paid to), CIP (carriage and insurance paid to), DAT (delivered at terminal), DAP (delivered at place) and DDP (delivered duty paid) can be used regardless of whether or not there is maritime transport. The terms FAS (free alongside ship), FOB, CFR (cost and freight) and CIF are used when the point of shipment and destination are both ports.

The two terms DAP and DAT were added in Incoterms 2010 while four terms, namely, DAF (delivered at frontier), DES (delivered ex ship), DEQ (delivered ex quay) and DDU (delivered duty unpaid) were excluded. The drafting group of the 2010 version of the rules considered that the new terms DAT and DAP were drawn widely enough to cater for contracts where the parties intended delivery to occur on a ship or at a quay and hoping that the market would use DAP and DAT for DES and DEQ, respectively. DAP and DAT may be used irrespective of the agreed mode of transport. Under both these terms, delivery occurs at a named destination. According to DAT, the seller delivers when the goods, once unloaded from the arriving means of transport, are placed at the disposal of the buyer at a named terminal at the named port or place of destination. The terminal should be clearly specified as the risks up to that point are for the account of the seller. The DAP term has similarities with DAT save that the seller delivers when the goods are placed at the disposal of the buyer on the arriving means of transport ready for unloading at the named place of destination. Instead of the terminal, the place of destination is used as the point determining when the risk passes to the buyer. Thus, both DAP and DAT may be used irrespective of the mode of transport selected and may also be used where more than one mode of transport is employed.

Once the contract for sale is concluded and the parties have decided on who will take responsibility for the carriage of goods, one or several contract(s) for carriage has to be arranged. The contract for carriage is similar to any other bilateral contract concluded upon exchange of mutual promises between the shipper

* The terms are periodically revised to keep pace with the development of international trade, the last revision being made in 2010.

and the carrier (Girvin, 2011: p. 83). In common law jurisdictions, a contract for carriage is governed by a combination of common law, tort, bailment and statute. In civil law jurisdictions, a contract for carriage is usually subject to the relevant civil and commercial codes. In a contract for sale, the parties are free to determine their rights and obligations and distribute the risk based on party autonomy. However, in contracts for sea carriage the contractual freedom is curbed by application of mandatory convention regimes, such as the Hague/Hague–Visby or Hamburg Rules. These Rules were introduced to protect shippers with lower bargaining power and create a minimum liability regime for the carriage of goods by sea.

Currently, international carriage of goods commonly involves multimodal carriage using mixed contracts, which encompass both sea and land transport. The ocean carriage in most parts of the world is regulated by Hague/Hague–Visby Rules or an adapted version of these Rules. There is much less harmony in the rules applicable to road or rail carriage as it is generally governed under national or regional law. In Europe, road carriage is governed by the CMR,* while rail carriage is governed by CIM/COTIF.† Also, in Europe, there exist two viewpoints for multimodal contracts. The first is 'plurality of contracts' wherein a multimodal transport contract contains several contracts for different modes of transport; thus subjecting the contracting carrier to different liability rules depending on the leg of the transport. The hugely debated decision of the English Court of Appeal in *Quantum Corp. Inc. v. Plane Trucking Ltd.*‡ reflects the plurality of contracts view, that is, the liability of the contracting carrier for the inland leg of transport is governed by CMR. Thus, under English law, the extension of the maritime law to the inland leg would not be possible as the inland leg of the multimodal contract falls under the CMR. The other view is that multimodal transport is governed by a *sui generis* contract wherein the contract is not regulated by the existing conventions. Therefore, some European countries, such as Germany and the Netherlands, have enacted legislation on multimodal transport.

Pursuant to Article 1(1) of the Multimodal Convention, '"[i]nternational multimodal transport" means the carriage of goods by at least two different modes of transport on the basis of a multimodal transport contract from a place in one country at which the goods are taken in charge by the multimodal transport operator to a place designated for delivery situated in a different country. The operations of pick-up and delivery of goods carried out in the performance of a unimodal transport contract, as defined in such contract, shall

* Convention on the Contract for the International Carriage of Goods by Road, 19 May 1956, 399 U.N.T.S. 189, amended by the 1978 Protocol (CMR).

† Convention concerning International Carriage by Rail 1980 (CIM-COTIF); the current version is the Uniform Rules Concerning the Contract for International Carriage of Goods by Rail, Appendix B to the Convention, amended by the 1999 Protocol.

‡ [2002] 2 Lloyd's Rep. 25.

not be considered as international multimodal transport'. Article 1(3) of the Convention stipulates that a '"[m]ultimodal transport contract" means a contract whereby a multimodal transport operator undertakes, against payment of freight, to perform or to procure the performance of international multimodal transport'.

The very concept of multimodal transport depends on the successful integration of various transport modes to provide a smooth delivery of goods from door to door. In multimodal transport, the carrier's role is fulfilled by a MTO who is responsible for undertaking a carriage from door to door under a single contract. The MTO unlike the carrier might not own any transport means and might not even be a carrier, but will only make arrangements and organise the transportation with various carriers, be it a shipowning company, a railway or a road carrier. The legal status of the MTO is crucial for determining its liability.

The Multimodal Convention introduced the concept of the MTO in Article 1(2) as 'any person who on his own behalf or through another person acting on his behalf concludes a multimodal transport contract and who acts as a principal, not as an agent or on behalf of the consignor or of the carriers participating in the multimodal transport operations, and who assumes responsibility for the performance of the contract'. Article 14 vests the MTO with responsibility 'for the goods from the time he takes the goods in his charge to the time of their delivery'. Section 4(1) of the UNCTAD/ICC Rules states that 'the MTO is responsible for the goods under these Rules from the time the MTO has taken the goods in his charge to the time of their delivery'.

A MTO can be a freight forwarder who not only fulfils the traditional function of being an intermediary between the carrier and the shipper but also provides all kinds of ancillary services to both parties. In multimodal transportation, a shipper usually contacts the freight forwarder, who then arranges the transport with one or several carriers on behalf of the shipper. The legal status of the freight forwarder in multimodal transportation is of particular interest, since they may act as an agent or as a principal. It is not always easy to draw a precise line between these two roles and the courts have to determine that on a case-by-case basis. To determine the legal status of the forwarder and whether it differs from a carrier, the court will have to examine the contract, since the substance of their obligations and also the remuneration agreed with the shipper can be of assistance (Hill, 1972: p. 16; Girvin, 2011: pp. 46–48). There is no mandatory convention governing freight forwarders' activity.*

* FIATA adopted its Model Rules for Freight Forwarding Services in 1996. The Rules are non-binding and apply only when they are incorporated into a contract by referring to the FIATA Model Rules for Freight Forwarding Services. Other well-known rules include The British International Freight Association (BIFA) Standard Trading Conditions, 2005A Edition (England and Wales) and the General Conditions of the Nordic Association of Freight Forwarders (NSAB) 2000.

Major loss or damage to goods occurs not when they are carried but at intermediate points during discharge, storage or transhipment. These intermediate points are generally managed by port and terminal operators, stevedores, warehousemen and so on, who act as the servants and agents of the carrier to fulfil the obligations under the contract of carriage. Most of the existing conventions do not cover these intermediate points and these servants or agents are not a party to the contract between the shipper and the carrier. In case of loss or damage to the goods when they are in the custody of these servants and agents, the question that arises is what would be the basis of liability and whether such liability is subject to any mandatory international regime. The other question is whether these servants and agents can invoke defences available to the carrier and limit their liability in accordance with one of the existing conventions. The Hague–Visby Rules introduced Article IV *bis*, according to which a category of carrier's servants and agents is entitled to invoke the defences and limits of liability available to the carrier, which is popularly known as the 'Himalaya' protection.* However, this provision does not apply to independent contractors, such as port and terminal operators or to stevedores.

DOCUMENTATION IN MULTIMODAL TRANSPORT

In an international sale involving the carriage of goods, a range of documents are issued both in the country of export and import. The documents issued before and at the time of shipment generally include various forms of transport documents, invoices, certificates and so on. This discussion only focuses on the transport documents, which can be categorised as negotiable and non-negotiable. The negotiable documents are prevalent in sea carriage, while the non-negotiable documents are inherent in other modes, such as road, rail and air. A bill of lading, which is a negotiable document used in sea carriage, fulfils three functions, namely, receipt for the goods, evidence of the contract and document of title. Non-negotiable documents, such as sea waybill, consignment note and air waybill only fulfil the function of receipt for the goods and evidence of the contract. The traditional role of the bill of lading in cross-border trade is unique with its three functions as it being the document of title to the goods gives control of the goods to the holder of the bill.

In multimodal transport, the popularly used documents are multimodal bills of lading, freight forwarders' bills of lading and combined transport bills of lading.† Pursuant to Rule 2 of the UNCTAD/ICC Rules, 'multimodal transport

* A Himalaya clause is a contractual provision expressed to be for the benefit of a third party who is not a party to the contract. The clause takes its name from a decision of the English Court of Appeal in the case of *Adler* v. *Dickson* (The Himalaya) [1954] 3 All ER 397.

† A sample copy of the negotiable multimodal transport bill of lading code-named MULTIDOC 95 issued by BIMCO, subject to the UNCTAD/ICC Rules for Multimodal Transport Documents is appended at the end of this chapter (Appendix, Figure 10.1).

document includes negotiable, non-negotiable transport documents as well as the case where the paper document has been replaced by electronic data interchange messages'. According to Article 5 of the Multimodal Convention 'when the goods are taken in charge by the multimodal transport operator, he shall issue a multimodal transport document which, at the option of the consignor, shall be in either negotiable or non-negotiable form'. Thus, the MTO assumes responsibility for the performance of the contract. It is common for freight forwarders to issue multimodal bills of lading. A freight forwarder may also issue a 'house' bill to the shipper, and then one of the carriers issues a bill naming the freight forwarder as the shipper or consignee. In such a case, each document should be scrutinised separately, but in the context of the overall contractual scheme.

Multimodal transport documents also fulfil the requirements of letters of credit. Article 19 of UCP 600 applies to cases where the credit calls for a multimodal transport document. It requires that one person shall be liable for the whole transit. The multimodal transport document should indicate the name of the carrier and be signed by the carrier or a named agent for or on behalf of the carrier, or the master or a named agent for or on behalf of the master.

In recent years, the desirability to replace paper-based transport documentation with electronic methods is remarkably noticeable in the international trading community. Currently, there exist both government and private initiatives in various parts of the world, based on registry and token models that have met with limited success in allowing parties to use electronic records in international trade transactions. In order to benefit the promotion of electronic communications in international trade, the United Nations Commission on International Trade Law (UNCITRAL) Working Group IV on Electronic Commerce recently finalised a model law on electronic transferable records.* It is envisaged that new harmonised e-commerce rules will allow the transport industry to embrace electronic records in the near future.

MULTIMODAL PROVISIONS IN VARIOUS TRANSPORT CONVENTIONS

Carriage by road

The CMR deals with international road carriage that has been widely adopted by countries mostly within Europe and the Middle East. The CMR provides that where a road vehicle containing goods is carried over part of the journey by sea, rail, inland waterways or air, without the goods being unloaded from the vehicle, the CMR applies to the entire transport operation, unless it is proved that the

* The UNCITRAL Working Group IV Reports and the Draft Model Law on Electronic Transferable Records, A/CN.9/WG.IV/WP.139, A/CN.9/WG.IV/WP.137/Add.1 and A/CN.9/WG.IV/WP.137/Add.2 are available at http://www.uncitral.org/uncitral/en/commission/working_groups/4Electronic_Commerce.html.

loss was not caused by the act or omission of the road carrier 'but by some event which could only have occurred in the course of and by reason of the carriage by that other means of transport'.* So, in case of an unattributed loss or where loss occurs during a non-road leg but due to the fault of the road carrier, the CMR applies to the entire carriage including non-road legs.

Carriage by rail

The CIM-COTIF deals with international rail carriage, which has been adopted by countries mostly within Europe and some in North Africa and the Middle East. Article 1(4) of the CIM 1999 provides that '[w]hen international carriage being the subject of a single contract of carriage includes carriage by sea or transfrontier carriage by inland waterway as a supplement to carriage by rail, these Uniform Rules shall apply if the carriage by sea or inland waterway is performed on services included in the list of services provided for in article 24(1) of the Convention'. Such a listing is, however, not required where the supplemental road or inland waterway carriage is national.† In addition, in relation to rail traffic on the listed routes, superadded grounds of exemption from liability are available.‡

Carriage by air

The Montreal Convention§ applies to 'all international carriage of persons, baggage or cargo performed by aircraft for reward'.¶ One of the major differences between the Montreal Convention and other mentioned transport conventions is that it applies to an actual carriage in contrast to contractual carriage. One of the important provisions related to its application to multimodal transport is found in Article 18(4) which provides that '[t]he period of the carriage by air does not extend to any carriage by land, by sea or by inland waterway performed outside an airport. If, however, such carriage takes place in the performance of a contract for carriage by air, for the purpose of loading, delivery or transhipment, any damage is presumed, subject to proof to the contrary, to have been the result of an event which took place during the carriage by air. If a carrier, without the consent of the consignor, substitutes carriage by another mode of transport for the whole or part of a carriage intended by the agreement between the parties to be carriage by air, such carriage by another mode of transport is deemed to be within the period of carriage by air'.

* See CMR, Article 2(1).

† See CIM, Article 1(3) which provides that when international carriage being the subject of a single contract includes carriage by road or inland waterway in internal traffic of a member state as a supplement to transfrontier carriage by rail, these uniform rules shall apply.

‡ See CIM, Article 38.

§ Convention for the Unification of Certain Rules for International Carriage by Air, 2242 U.N.T.S. 309; S. Treaty Doc. No. 106-45 (2000) (Montreal Convention).

¶ See Montreal Convention, Article 1(1).

The foregoing provision clarifies that the Montreal Convention does not really extend its application to other modes of transport performed outside an airport. Thus, the only situation when it does apply is when such carriage takes place for the performance of the contract of carriage for the purposes of loading, delivery or transhipment. In that case, there is a presumption that the damage occurred during the air carriage. Such an 'air plus' arrangement has a restriction application and does not purport to expand to other modes.

Carriage wholly or partly by sea

In 2008, the UN General Assembly adopted the Rotterdam Rules* with an aim to create a modern and uniform law concerning the international carriage of goods by sea. In sharp contrast with some of the existing international maritime transport conventions, the application of the Rotterdam Rules is contractual which is defined by the contract of carriage itself.† If the contract covers land carriage preceding the loading of the vessel and land carriage subsequent to the unloading of the vessel, then the Rules also cover such carriage. But if the contract covers only the maritime leg of a multimodal movement, then that is all that the Rules cover. In other words, if a contract of carriage provides for a shipment from one port to another port, then the Rules' coverage is simply 'port to port'. But if a contract of carriage provides for a shipment from the shipper's manufacturing plant to the consignee's warehouse, then the Rules' coverage is 'door to door'.‡

The door-to-door coverage of the Rotterdam Rules is different from traditional multimodal coverage. Ideally, in a multimodal regime, the contract of carriage should provide for any two (or more) modes of carriage§ and could also govern a shipment involving only road and rail transport. The Rules, in contrast, require a maritime leg¶ and create a modified door-to-door solution known as a 'maritime

* United Nations Convention on Contracts for the International Carriage of Goods Wholly or Partly by Sea, 23 September 2009, United Nations Publication, Sales No. E.09.V.9 (Rotterdam Rules), available online at http://www.uncitral.org/pdf/english/texts/transport/rotterdam_rules/09-85608_Ebook. The Rules have so far received 25 signatures and 3 ratifications; see 'Status of the Rotterdam Rules' http://www.uncitral.org/uncitral/en/uncitral_texts/transport_goods/rotterdam_status.html. Pursuant to Article 94, the convention requires ratification or accession by at least 20 states to enter into force.

† Rotterdam Rules, Article 1(1).

‡ Similarly, the Rotterdam Rules' coverage may be 'door to port' or 'port to door', depending on the scope of the contract. See Rotterdam Rules, Article 11.

§ Article 1(1) of the Multimodal Convention defines 'international multimodal transport' as 'the carriage of goods by at least two different modes of transport on the basis of a multimodal transport contract from a place in one country at which the goods are taken in charge by the multimodal transport operator to a place designated for delivery situated in a different country'.

¶ Rotterdam Rules, Article 1(1), 'The contract shall provide for carriage by sea and may provide for carriage by other modes of transport in addition to the sea carriage'.

plus' or 'limited network' regime.* This means that when there is a through bill of lading covering an international multimodal shipment, and one of the legs of the journey is by sea, then, as between the parties to the contract, the convention's terms, including its liability terms, apply, regardless of whether the damage occurred on the ocean leg or during the inland carriage.† The obligations relating to the care of the cargo apply throughout the period of responsibility of the carrier‡ whereas the obligations relating to the seaworthiness of the ship apply only to the voyage by sea.§

The Rotterdam Rules, to avoid conflict with the CMR and the CIM-COTIF, provide in Article 26 that, if it can be proved that the damage occurred during land transport that would have been subject to a mandatorily applicable international convention, then that land convention will apply, otherwise these Rules will apply. As an illustration, the CMR regime in relation to liability will apply to the carrier if a separate contract was entered into between the carrier and the cargo interest in relation to the road carriage before the goods were loaded on the ship or after their discharge. In such case, that contract would be governed by the CMR, if the loss occurred during the road transit before loading or after discharge. Article 26 only applies to unimodal provisions on the liability of the carrier, limitation of liability and time for suit. Other provisions of the Rotterdam Rules, such as those on jurisdiction, shipper's liability, transport documents and rights of control, will continue to apply concurrently with those of other conventions. Article 82 of the Rules prevents conflict of conventions by giving precedence to those conventions already in force at the date the Rules enter into force.

LIABILITY REGIMES IN MULTIMODAL TRANSPORT

Multimodal transport has its advantages but there exist certain legal impediments. The first challenge is the localisation of loss or damage as it is not always straightforward to determine at which stage of transport the loss or damage occurred. In the container trade, loss is often concealed as the container is sealed upon receipt and is not opened until delivery. Even if the damage is identified, it may occur gradually or span two legs. The other major challenge is delay as it can be more problematic to detect the stage of transport and the time and location when delay has occurred. The delivery time of goods in multimodal transport is crucial largely due to the pressure of just-in-time delivery. The carrier has to comply with the strictures of a complicated logistics chain and the slightest delay on one transport leg may produce a series of further delays resulting in sufficient delay for the claimant. Therefore, the question of localisation of loss, damage and delay remains largely problematic since it has implications regarding which legal regime should apply and raises the issues of time-bar and limitation of liability.

* Rotterdam Rules, Articles 5–7.
† Rotterdam Rules, Articles 11 and 12.
‡ Rotterdam Rules, Article 13(1).
§ Rotterdam Rules, Article 14.

The liability of the MTO is of utmost significance in multimodal transport. As previously discussed, frequently the MTO engages other intermediaries to perform certain stages of transportation and is liable for them. Thus, in case of loss of or damage to the goods, the MTO needs to have recourse from the sub-carriers based on the respective conventions. The lack of a single instrument applicable to all stages aggravates the situation both for the MTO and a cargo owner. Since damage in the container trade is mostly unlocalised, there is no predictability regarding the applicable liability regime and the limitation amounts. The primary question in this regard is whether the liability should be uniform for all stages of the multimodal transport or separate for each part of the different modes of transport used. There are two main approaches to the liability of the MTO: the 'uniform liability system' and the 'network liability system'. Both systems are theoretical concepts and when applied in practice there are certain impediments pertinent to each system. Hence, they are mostly used in modified form.

Uniform liability system

The uniform liability system involves the application of a single regime throughout the transport modes regardless of the localisation of damage or loss. During the entire carriage, the MTO will be responsible to the cargo interests under a contract for the multimodal carriage of goods. This contract will be subject to specific rules, which may be different from the unimodal rules applicable to each separate leg of the transport (De Wit, 1995: p. 138). The advantage of a uniform system is that one regime applies both with respect to localised and unlocalised damage. Hence, it is simple, transparent and predictable to all parties, as they already know from the onset what would be the basis of liability and the limitation amounts. However, it presents difficulties for carriers, given that their rights of recourse against the sub-contractors will be governed by the relevant unimodal regime. The liability regime introduced in the Multimodal Convention was based on the uniform liability approach where the MTO was responsible from taking the goods in charge until their delivery.*

Network liability system

In the network liability system, where the damage is localised, it will be governed by the regime applicable to that particular stage. Thus, if the damage occurred during the road leg in Europe, the CMR would apply. The laws governing different parts of transport used would apply one after the other and then operate in the framework of multimodal transport, just as they would in unimodal transport. As a result, conflicts with mandatory conventions would be minimised and the MTOs would be entitled to take recourse from their sub-contractors.

The network system exists in its pure and modified form. It has been argued that the pure network liability system causes enormous problems and disadvantages with

* Multimodal Convention, Articles 14–21.

respect to its practical application, especially in unlocalised loss or damage, gradually occurring loss or damage and delay in delivery as the outcome can be unpredictable. Even when the damage is localised, that stage might not be covered by any transport convention. Hence, the network liability might be insufficient and create gaps when there is no carrier liability at all (De Wit, 1995: p. 139). This might happen when damage or loss occurs between the modes and none of the conventions cover that period. As noted earlier, such loss or damage is a frequent occurrence. From the MTO's point of view, a pure network system has the advantage that its liability to the shipper should correspond with the liability of any performing carrier. However, from the shipper's point of view, the network system has certain pitfalls. In a concealed damage scenario, localisation of the loss may be nearly impossible. This also concerns gradual damage during the whole journey or the spreading of damage over several stages, to say nothing of the cases of delay. Thus, if there are no contractually agreed rules, the parties cannot assess their risks in advance, which might lead to higher insurance costs. The pure network system contravenes the purpose of multimodal transport by cutting the integrated transport into pieces. Such a system devised by lawyers with unimodal thinking fails to attend to the need of transport operators who integrate their services to provide a multimodal solution.

The industry has tested both regimes and practice has shown that a modified network liability approach is more efficient than the uniform system. However, the existence of the different limitation amounts in various transport conventions might encourage the claimant to prove that the damage occurred at a particular stage of transport to claim the highest applicable limitation amount (Grönfors, 1967). It is notable that the network liability system has been introduced in some domestic laws on multimodal transport such as in the Netherlands and Germany.*

CONCLUSION

Multimodal transport is currently performed on the basis of multiple transport documents to cover various modes, or based on a single transport document issued by a MTO. If there is cargo loss, damage or delay, then the shipper claims damages from the carrier or transport operator. On the other hand, the carrier or operator, except in the situation where it is liable without question, seeks to deny, exclude or limit liability for the loss, damage or delay. Although the claim itself and its handling may present various complexities than the objective of the claim in a practical context, the lack of a single comprehensive convention regulating multimodal transport makes the situation even more complex. A patchwork of various liability regimes reduces predictability for all parties concerned. A comprehensive insurance is used in some transport modes while it is less common in certain other modes of transport. International efforts have been undertaken several times but these have not met with sufficient success. It is difficult to fathom precisely all the reasons why these efforts have failed to gain international acceptance. Technological progress and electronic

* See the Dutch Civil Code and the German Commercial Code.

means of communication have moved far ahead of the legal regimes. This strongly indicates the need for a single instrument to apply throughout the multimodal chain to facilitate legal certainty and predictability for the parties involved.

REFERENCES

Books and articles

Bridge, M. (1999). *The International Sale of Goods. Law and Practice*. Oxford: OUP.

De Wit, R. (1995). *Multimodal Transport: Carrier Liability and Documentation*. London: LLP.

Girvin, S. (2011). *Carriage of Goods by Sea*. 2nd edn. Oxford: OUP.

Grönfors, K. (1967). Container transport and the Hague Rules. *Journal of International Business Law*. 298.

Hill, D. J. (1972). *Freight Forwarders*. London: Stevens & Sons.

Ramberg, J. (2011). *International Commercial Transactions*. 4th edn. Stockholm: Nordsteds Juridik AB.

International instruments

Convention concerning International Carriage by Rail 1980 (CIM-COTIF).

Convention for the Unification of Certain Rules for International Carriage by Air, 2242 U.N.T.S. 309; S. Treaty Doc. No. 106-45 (2000) (Montreal Convention).

Convention on the Contract for the International Carriage of Goods by Road, 19 May 1956, 399 U.N.T.S. 189, amended by the 1978 Protocol (CMR).

Draft Convention on Combined Transport, 1969.

Draft Convention on the International Combined Transport of Goods (TCM).

Draft Model Law on Electronic Transferable Records A/CN.9/WG.IV/WP.139, A/CN.9/WG.IV/WP.139/Add.1 and A/CN.9/WG.IV/WP.139/Add.2.

International Convention for the Unification of Certain Rules Relating to Bills of Lading, 25 August 1924, 120 L.N.T.S. 155 (Hague Rules).

Protocol to Amend the International Convention for the Unification of Certain Rules Relating to Bills of Lading 1924, 23 February 1968, 1421 U.N.T.S. 121 (Hague–Visby Rules).

UNCTAD/ICC Rules for Multimodal Transport Documents, ICC publication No. 481, ICC, Paris, 1992 (UNCTAD/ICC Rules 1992).

Uniform Rules for a Combined Transport Document, ICC publication No. 298, ICC, Paris, 1975 (ICC Rules 1975).

United Nations Convention on Contracts for the International Carriage of Goods Wholly or Partly by Sea, 23 September 2009 (Rotterdam Rules).

United Nations Convention on International Multimodal Transport of Goods, 24 May 1980, U.N. Doc. TD/MT/CONF/16 (1980) (Multimodal Convention).

United Nations Convention on the Carriage of Goods by Sea, 31 March 1978, 17 I.L.M. 608 (Hamburg Rules).

Cases

Adler v. Dickson (The Himalaya) [1954] 3 All ER 397

Quantum Corp. Inc. v. Plane Trucking Ltd [2002] 2 Lloyd's Rep. 25

APPENDIX

First published 1995, revised 2016

BIMCO

MULTIDOC 2016

NEGOTIABLE MULTIMODAL TRANSPORT BILL OF LADING

Subject to the UNCTAD/ICC Rules for Multimodal Transport Documents, (ICC Publication No. 481)

Page 1

Consignor	MT Doc. No.	Reference No.
Consigned to order of	Notify address	Vessel
Place of receipt	Port of loading	
Place of delivery	Port of discharge	

Marks and Nos.	Quantity and description of goods	Gross weight, kg	Measurement, m^3

Particulars above declared by Consignor

Freight and charges

Freight payable at

Consignor's declared value of

subject to payment of above extra charge.

Note: The Merchant's attention is called to the fact that according to Clauses 10-12 and 26 of this MT Bill of Lading, the liability of the MTO is, in most cases, limited in respect of loss of or damage to the goods.

RECEIVED the goods in apparent good order and condition and, as far as ascertained by reasonable means of checking, as specified above unless otherwise stated.

The MTO, in accordance with and to the extent of the provisions contained in this MT Bill of Lading, and with liberty to sub-contract, undertakes to perform and/or in its own name to procure performance of the multimodal transport and the delivery of the goods, including all services related thereto, from the place and time of taking the goods in charge to the place and time of delivery and accepts responsibility for such transport and such services.

One of the MT Bills of Lading must be surrendered duly endorsed in exchange for the goods or delivery order.

IN WITNESS whereof MT Bill(s) of Lading has/have been signed in the number indicated below, one of which being accomplished the other(s) to be void.

Place and date of issue	Number of original MT Bills of Lading

MTO as Carrier:...(insert name)

Signature:..(MTO as Carrier*/Master*/Agent*)
**Delete as appropriate*

If signed by an Agent indicate with a tick ☑ whether for and on behalf of:

☐ Master; or

☐ MTO as Carrier

Agent..(insert name)

v. 1.1. Dated 30 May 2016. Clause 24 (General Average) updated to refer to York-Antwerp Rules 2016.

Figure 10.1 Multimodal bill of lading. (From BIMCO.)

MULTIDOC 2016
NEGOTIABLE MULTIMODAL TRANSPORT BILL OF LADING
Page 2

I. GENERAL PROVISIONS
1. Applicability
The provisions of this Contract shall apply irrespective of whether there is a unimodal or a Multimodal Transport Contract involving one or several modes of transport.
2. Definitions
"*Multimodal Transport Contract*" means a single Contract for the carriage of Goods by at least two different modes of transport. "*Multimodal Transport Bill of Lading*" (MT Bill of Lading) means this document evidencing a Multimodal Transport Contract and which can be replaced by electronic data interchange messages insofar as permitted by applicable law and is issued in a negotiable form.
"*Multimodal Transport Operator*" (MTO) means the person named on the face hereof who concludes a Multimodal Transport Contract and assumes responsibility for the performance thereof as a Carrier.
"*Carrier*" means the person who actually performs or undertakes to perform the carriage, or part thereof, whether it is identical with the Multimodal Transport Operator or not.
"*Merchant*" includes the Shipper, the Receiver, the Consignor, the Consignee, the holder of this MT Bill of Lading and the owner of the Goods.
"*Consignor*" means the person who concludes the Multimodal Transport Contract with the Multimodal Transport Operator.
"*Consignee*" means the person entitled to receive the Goods from the Multimodal Transport Operator.
"*Taken in charge*" means that the Goods have been handed over to and accepted for carriage by the MTO.
"*Delivery*" means
(i) the handing over of the Goods to the Consignee; or
(ii) the placing of the Goods at the disposal of the Consignee in accordance with the Multimodal Transport Contract or with
the law or usage of the particular trade applicable at the place of delivery; or
(iii) the handing over of the Goods to an authority or other third party to whom, pursuant to the law or regulations applicable at the place of delivery, the Goods must be handed over.
"*Special Drawing Rights*" (SDR) means the unit of account as defined by the International Monetary Fund.
"*Goods*" means any property including live animals as well as containers, pallets or similar articles of transport or packaging not supplied by the MTO, irrespective of whether such property is to be or is carried on or under deck.
3. MTO's Tariff
The terms of the MTO's applicable tariff at the date of shipment are incorporated herein. Copies of the relevant provisions of the applicable tariff are available from the MTO upon request. In the case of inconsistency between this MT Bill of Lading and the applicable tariff, this MT Bill of Lading shall prevail.
4. Time Bar
The MTO shall, unless otherwise expressly agreed, be discharged of all liability under this MT Bill of Lading unless suit is brought within nine months after:
(i) the Delivery of the Goods; or
(ii) the date when the Goods should have been delivered; or
(iii) the date when, in accordance with sub-clause 10 (e) failure to deliver the Goods would give the Consignee the right to treat the Goods as lost.
5. Law and Jurisdiction
Disputes arising under this MT Bill of Lading shall be determined by the courts and in accordance with the law at the place where the MTO has its principal place of business.

II. PERFORMANCE OF THE CONTRACT
6. Methods and Routes of Transportation
(a) The MTO is entitled to perform the transport in any reasonable manner and by any reasonable means, methods and routes.
(b) In accordance herewith, for instance, in the event of carriage by sea, vessels may sail with or without pilots, undergo repairs, adjust equipment, drydock and tow vessels in all situations.
7. Optional Stowage
(a) Goods may be stowed by the MTO by means of containers, trailers, transportable tanks, flats, pallets, or similar articles of transport used to consolidate Goods.
(b) Containers, trailers, transportable tanks and covered flats, whether stowed by the MTO or received by him in a stowed condition, may be carried on or under deck without notice to the Merchant.
8. Delivery of the Goods to the Consignee
The MTO undertakes to perform or to procure the performance of all acts necessary to ensure Delivery of the Goods:
(i) when the MT Bill of Lading has been issued in a negotiable form "to bearer", to the person surrendering one original of the document; or
(ii) when the MT Bill of Lading has been issued in a negotiable form "to order", to the person surrendering one original of the document duly endorsed; or
(iii) when the MT Bill of Lading has been issued in a negotiable form to a named person, to that person upon proof of its identity and surrender of one original document; if such document has been transferred "to order" or in blank, the provisions of (ii) above apply.
9. Hindrances, etc. Affecting Performance
(a) The MTO shall use reasonable endeavours to complete the transport and to deliver the Goods at the place designated for Delivery.
(b) If at any time the performance of the Contract as evidenced by this MT Bill of Lading is or will be affected by any hindrance, risk, delay, difficulty or disadvantage of whatsoever kind and if by virtue of sub-clause 9 (a) the MTO has no duty to complete the performance of the Contract, the MTO (whether or not the transport is commenced) may elect to
(i) treat the performance of this Contract as terminated and place the Goods at the Merchant's disposal at any place which the MTO shall deem safe and convenient; or
(ii) deliver the Goods at the place designated for Delivery.
(c) If the Goods are not taken Delivery of by the Merchant within a reasonable time after the MTO has called upon him to take Delivery, the MTO shall be at liberty to put the Goods in safe custody on behalf of the Merchant at the latter's risk and expense.
(d) In any event the MTO shall be entitled to full freight for Goods received for transportation and additional compensation for extra costs resulting from the circumstances referred to above.

III. LIABILITY OF THE MTO
10. Basis of Liability
(a) The responsibility of the MTO for the Goods under this Contract covers the period from the time the MTO has taken the Goods into its charge to the time of their Delivery.
(b) Subject to the defences set forth in Clauses 11 and 12, the MTO shall be liable for loss of or damage to the Goods as well as for delay in Delivery, if the occurrence which caused the loss, damage or delay in Delivery took place while the Goods were in its charge as defined in sub-clause 10 (a), unless the MTO proves that no fault or neglect of its own, its servants or agents or any other person referred to in sub-clause 10 (c) has caused or contributed to the loss, damage or delay in Delivery.
However, the MTO shall only be liable for loss following from delay in Delivery if the Consignor has made a written declaration of interest in timely Delivery which has been accepted in writing by the MTO.
(c) The MTO shall be responsible for the acts and omissions of its servants or agents, when any such servant or agent is acting within the scope of its employment, or of any other person of whose services it makes use for the performance of the Contract, as if such acts and omissions were its own.
(d) Delay in Delivery occurs when the Goods have not been delivered within the time expressly agreed upon or, in the absence of such agreement, within the time which it would be reasonable to require of a diligent MTO, having regard to the circumstances of the case.
(e) If the Goods have not been delivered within ninety (90) consecutive days following the date of Delivery determined according to Clause 10 (d) above, the claimant may, in the absence of evidence to the contrary, treat the Goods as lost.
11. Defences for Carriage by Sea or Inland Waterways
Notwithstanding the provisions of Clause 10 (b), the MTO shall not be responsible for loss, damage or delay in Delivery with respect to Goods carried by sea or inland waterways when such loss, damage or delay during such carriage results from:
(i) act, neglect or default of the master, mariner, pilot or the servants of the Carrier in the navigation or in the management of the vessel;
(ii) fire, unless caused by the actual fault or privity of the Carrier;
(iii) the causes listed in the Hague-Visby Rules article 4.2 (c) to (p);
however, always provided that whenever loss or damage has resulted from unseaworthiness of the vessel, the MTO can prove that due diligence has been exercised to make the vessel seaworthy at the commencement of the voyage.
12. Limitation of Liability
(a) Unless the nature and value of the Goods have been declared by the Consignor before the Goods have been taken in charge by the MTO and inserted in the MT Bill of Lading, the MTO shall in no event be or become liable for any loss of or damage to the Goods in an amount exceeding:
(i) when the Carriage of Goods by Sea Act of the United States of America, 1936 (US COGSA) applies USD 500 per package or customary freight unit; or
(ii) when any other law applies, the equivalent of 666.67 SDR per package or unit or two SDR per kilogramme of gross weight of the Goods lost or damaged, whichever is the higher.
(b) Where a container, pallet or similar article of transport is loaded with more than one package or unit, the packages or other shipping units enumerated in the MT Bill of Lading as packed in such article of transport are deemed packages or shipping units. Except as aforesaid, such article of transport shall be considered the package or unit.
(c) Notwithstanding the above-mentioned provisions, if the Multimodal Transport does not, according to the Contract, include carriage of Goods by sea or by inland waterways, the liability of the MTO shall be limited to an amount not exceeding 8.33 SDR per kilogramme of gross weight of the Goods lost or damaged.
(d) In any case, when the loss of or damage to the Goods occurred during one particular stage of the Multimodal Transport, in respect of which an applicable international convention or mandatory national law would have provided another limit of liability if a separate contract of carriage had been made for that particular stage of transport, then the limit of the MTO's liability for such loss or damage shall be determined by reference to the provisions of such convention or mandatory national law.
(e) If the MTO is liable in respect of loss following from delay in Delivery, or consequential loss or damage other than loss of or damage to the Goods, the liability of the MTO shall be limited to an amount not exceeding the equivalent of the freight under the Multimodal Transport Contract for the Multimodal Transport.
(f) The aggregate liability of the MTO shall not exceed the limits of liability for total loss of the Goods.
(g) The MTO is not entitled to the benefit of the limitation of liability if it is proved that the loss, damage or delay in Delivery resulted from a personal act or omission of the MTO done with the intent to cause such loss, damage or delay, or recklessly and with knowledge that such loss, damage or delay would probably result.
13. Assessment of Compensation
(a) Assessment of compensation for loss of or damage to the Goods shall be made by reference to the value of such Goods at the place and time they are delivered to the Consignee or at the place and time when, in accordance with the Multimodal Transport Contract, they should have been so delivered.
(b) The value of the Goods shall be determined according to the current commodity exchange price or, if there is no such price, according to the current market price or, if there is no commodity exchange price or current market price, by reference to the normal value of Goods of the same kind and quality.
14. Notice of loss of or Damage to the Goods
(a) Unless notice of loss of or damage to the Goods, specifying the general nature of such loss or damage, is given in writing by the Consignee to the MTO when the Goods are handed over to the Consignee, such handing over is prima facie evidence of the Delivery by the MTO of the Goods as described in the MT Bill of Lading.
(b) Where the loss or damage is not apparent, the same prima facie effect shall apply if notice in writing is not given within six consecutive days after the day when the Goods were handed over to the Consignee.
15. Defences and Limits for the MTO, Servants, etc.
The provisions of this Contract apply to all claims against the MTO relating to the performance of the Multimodal Transport Contract, whether the claim be founded in contract or in tort.
16. International Group of P&I Clubs/BIMCO Himalaya Clause for bills of lading and other contracts 2014
(a) For the purposes of this contract, the term "Servant" shall include the owners, managers, and operators of vessels (other than the Carrier); underlying carriers; stevedores and terminal operators; and any direct or indirect servant, agent, or subcontractor (including their own subcontractors), or any other party employed by or on behalf of the Carrier, or whose services or equipment have been used to perform this contract whether in direct contractual privity with the Carrier or not.
(b) It is hereby expressly agreed that no Servant shall in any circumstances whatsoever be under any liability whatsoever to the Merchant or other party to this contract (hereinafter termed "Merchant") for any loss, damage or delay of whatsoever kind arising or resulting directly or indirectly from any act, neglect or default on the Servant's part while acting in the course of or in connection with the performance of this contract.
(c) Without prejudice to the generality of the foregoing provisions in this clause, every exemption, limitation, condition and liberty contained herein (other than Art III Rule 8 of the Hague/Hague-Visby Rules if incorporated herein) and every right, exemption from liability, defence and immunity of whatsoever nature applicable to the carrier or to which the carrier is entitled hereunder including the right to enforce any jurisdiction or arbitration provision contained herein shall also be available and shall extend to every such Servant of the carrier, who shall be entitled to enforce the same against the Merchant.
(d) (i) The Merchant undertakes that no claim or allegation whether arising in contract, bailment, tort or otherwise shall be made against any Servant of the carrier which imposes or attempts to impose upon any of them or any vessel owned or chartered by any of them any liability whatsoever in connection with this contract whether or not arising out of negligence on the part of such Servant. The Servant shall also be entitled to enforce the foregoing covenant against the Merchant; and
(ii) The Merchant undertakes that if any such claim or allegation should nevertheless be made, it will indemnify the carrier against all consequences thereof.
(e) For the purpose of sub-paragraphs (a)-(d) of this clause the Carrier is or shall be deemed to be acting as agent or trustee on behalf of and for the benefit of all persons mentioned in sub-clause (a) above who are its Servant and all such persons shall to this extent be or be deemed to be parties to this contract.

IV. DESCRIPTION OF GOODS
17. MTO's Responsibility
The information in the MT Bill of Lading shall be prima facie evidence of the taking in charge by the MTO of the Goods as described by such information unless a contrary indication, such as "shipper's weight, load and count", "shipper-packed container" or similar expressions, have been made in the printed text or superimposed on the document. Proof to the contrary shall not be admissible when the MT Bill of Lading has been transferred, or the equivalent electronic data interchange message has been transmitted to and acknowledged by the Consignee who in good faith has relied and acted thereon.
18. Consignor's Responsibility
(a) The Consignor shall be deemed to have guaranteed to the MTO the accuracy, at the time the Goods were taken in charge by the MTO, of all particulars relating to the general nature of the Goods, their marks, number, weight, volume and quantity and, if applicable, to the dangerous character of the Goods as furnished by it or on its behalf for insertion in the MT Bill of
Lading.
(b) The Consignor shall indemnify the MTO for any loss or expense caused by inaccuracies in or inadequacies of the particulars referred to above.
(c) The right of the MTO to such indemnity shall in no way limit his liability under the Multimodal Transport Contract to any person other than the Consignor.
(d) The Consignor shall remain liable even if the MT Bill of Lading has been transferred by him.
19. Return of Containers
(a) Containers, pallets or similar articles of transport supplied by or on behalf of the MTO shall be returned to the MTO in the same order and condition as when handed over to the Merchant, normal wear and tear excepted, with interiors clean and within the time prescribed in the MTO's tariff or elsewhere.
(b) (i) The Consignor shall be liable for any loss of, damage to, or delay, including demurrage, of such articles, incurred during the period between handing over to the Consignor and return to the MTO for carriage.
(ii) The Consignor and the Consignee shall be jointly and severally liable for any loss of, damage to, or delay, including demurrage, of such articles, incurred during the period between handing over to the Consignee and return to the MTO.
20. Dangerous Goods
(a) The Consignor shall comply with all internationally recognised requirements and all rules which apply according to national law or by reason of international convention, relating to the carriage of Goods of a dangerous nature, and shall in any event inform the MTO in writing of the exact nature of the danger before Goods of a dangerous nature are taken in charge by the MTO and indicate to him, if need be, the precautions to be taken.
(b) If the Consignor fails to provide such information and the MTO is unaware of the dangerous nature of the Goods and the necessary precautions to be taken and if, at any time, they are deemed to be a hazard to life or property, they may at any place be unloaded, destroyed or rendered harmless, as circumstances may require, without compensation and the Consignor shall be liable for all loss, damage, delay or expenses arising out of their being taken in charge, or their carriage, or of any service incidental thereto.
The burden of proving that the MTO knew the exact nature of the danger constituted by the carriage of the said Goods shall rest upon the person entitled to the Goods.
(c) If any Goods shipped with the knowledge of the MTO as to their dangerous nature shall become a danger to the vessel or cargo, they may in like manner be landed at any place or destroyed or rendered innocuous by the MTO without liability on the part of the MTO except to General Average, if any.
21. Consignor-packed Containers, etc.
(a) If a container has not been filled, packed or stowed by the MTO, the MTO shall not be liable for any loss of or damage to its contents and the Consignor shall indemnify any loss or expense incurred by the MTO if such loss, damage or expense has been caused by:
(i) negligent filling, packing or stowing of the container;
(ii) the contents being unsuitable for carriage in container; or
(iii) the unsuitability or defective condition of the container unless the container has been supplied by the MTO and the unsuitability or defective condition would not have been apparent upon reasonable inspection at or prior to the time when the container was filled, packed or stowed.
(b) The provisions of sub-clause (a) of this Clause also apply with respect to trailers, transportable tanks, flats and pallets which have not been filled, packed or stowed by the MTO.
(c) The MTO does not accept liability for damage due to the unsuitability or defective condition of reefer equipment or trailers supplied by the Merchant.

V. FREIGHT AND LIEN
22. Freight
(a) Freight shall be deemed earned when the Goods have been taken into charge by the MTO and shall be paid in any event.
(b) The Merchant's attention is drawn to the stipulations concerning currency in which the freight and charges are to be paid, rate of exchange, devaluation and other contingencies relative to freight and charges in the relevant tariff conditions. If no such stipulation as to devaluation exists or is applicable the following provision shall apply:
If the currency in which freight and charges are quoted is devalued or revalued between the date of the freight agreement and the date when the freight and charges are paid, then all freight and charges shall be automatically and immediately changed in proportion to the extent of the devaluation or revaluation of the said currency. When the MTO has consented to payment in other currency than the above mentioned currency, then all freight and charges shall - subject to the preceding paragraph - be paid at the highest selling rate of exchange for banker's sight draft current on the day when such freight and charges are paid. If the banks are closed on the day when the freight is paid the rate to be used will be the one in force on the last day the banks were open.
(c) For the purpose of verifying the freight basis the MTO reserves the right to have the contents of containers, trailers or similar articles of transport inspected in order to ascertain the weight, measurement, value, or nature of the Goods. If on such inspection it is found that the declaration is not correct, it is agreed that a sum equal either to five times the difference between the correct freight and the freight charges or to double the correct freight less the freight charges, whichever sum is the smaller, shall be payable as liquidated damages to the MTO notwithstanding any other sum having been stated on this MT Bill of Lading as the freight payable.
(d) All dues, taxes and charges levied on the Goods and other expenses in connection therewith shall be paid by the Merchant.
23. Lien
The MTO shall have a lien on the Goods for any amount due under this Contract and for the costs of recovering the same, and may enforce such lien in any reasonable manner, including sale or disposal of the Goods.

VI. MISCELLANEOUS PROVISIONS
24. General Average
(a) General Average shall be adjusted at any port or place at the MTO's option, and to be settled according to the York-Antwerp Rules 2016, this covering all Goods, whether carried on or under deck. The New Jason Clause as approved by BIMCO to be considered as incorporated herein.
(b) Such security including a cash deposit as the MTO may deem sufficient to cover the estimated contribution of the Goods and any salvage and special charges thereon, shall, if required, be submitted to the MTO prior to Delivery of the Goods.
25. Both-to-Blame Collision Clause
The Both-to-Blame Collision Clause as adopted by BIMCO shall be considered incorporated herein.
26. U.S. Trade
In case the Contract evidenced by this MT Bill of Lading is subject to U.S COGSA, then the Provisions stated in said Act shall govern before loading and after discharge and throughout the entire time the Goods are in the Carrier's custody.

Figure 10.1 (Continued) Multimodal bill of lading. (From BIMCO.)

Part IV

ANALYSIS

Chapter 11

Modelling of intermodal systems

Jonas Flodén, Dries Meers and Cathy Macharis

COMPLEXITY OF DESIGNING INTERMODAL SYSTEMS

Intermodal transport systems are complex to manage and design. The descriptions of intermodal transport in the previous chapters show that, compared with unimodal transport, the former is often, and in many aspects, much more complex. Different transport modes are required to fulfil transport needs, and additional actors are involved in transport operations and planning. When designing and evaluating complex systems, modelling is an important part of the decision-making process. The models are tools that help decision makers make informed decisions about the transport system. Intermodal transport system models distinguish themselves from general transport models with the complexity of the modelling due to the larger number of actors involved and the inherent aspect of modal transfers.

Overlooking all potential design aspects without the help of models is a daunting task. For example, the terminal network configuration determines, to a large extent, the markets that intermodal transport can serve. Terminals should thus be located where they can serve a sufficiently large customer base and provide cost-competitive transport services in a large market area (Niérat, 1997) (see Figure 11.1). A low density of transhipment terminals limits the use of intermodal transport, particularly in remote areas, while a high density of terminals can result in competition between terminals, which can decrease the profitability of some terminals. The transport infrastructure network configuration is obviously related to the terminal network configuration. The transport network and the terminal design limit the transport capacity (e.g. the width and depth of a canal limits the size of barges that can be used, and the length of rail tracks at a terminal limits train length) and thus influence the profitability of potential services by reducing the possibilities for economies of scale.

Similarly, networks and operations can be designed in a multitude of ways. Figure 11.2 displays the five main intermodal network design types (Woxenius and Bärthel, 2008). The direct connection (D2) and shuttle trains (D3) represent the most basic and often most profitable design, while the more complex designs – such as the hierarchic network (D1), hub-and-spoke (D4) and transport corridor (D5) – allow for operations in regions with lower goods volumes, although

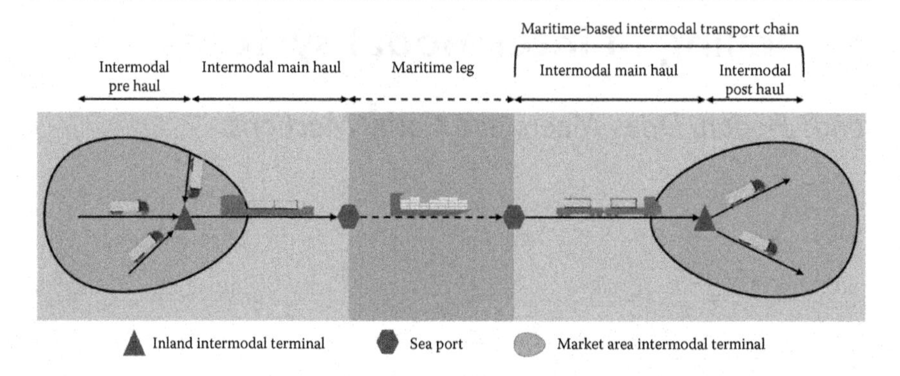

Figure 11.1 Intermodal transport chain (not to scale) depicting an intermodal terminal market area. (From Meers, D. et al., *Sustainable Logistics*, Emerald, Bingley, 2014.)

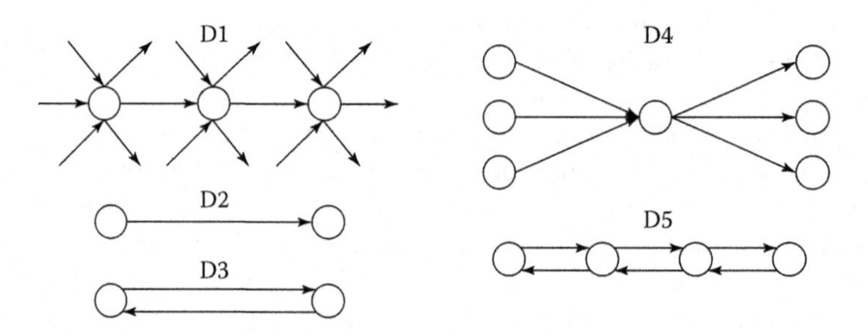

Figure 11.2 Intermodal network design types connecting two or more intermodal hubs. (From Woxenius, J. and Bärthel, F., *The Future of Intermodal Freight Transport: Operations, Design and Implementation*, Edward Elgar Publishing. Cheltenham, UK, 2008.)

transport times increase (Behrends and Flodén, 2012). The management of operations is further subjected to a number of design choices. Outsourcing is very common; for example, the train operations are outsourced to a rail operator, who in turn might lease their equipment. Similarly, terminals are commonly operated by a third-party logistics company. The chosen system design clearly impacts the cost and quality of the transport services, thereby impacting the transport buyers' preferences for a given alternative. Balancing all of these factors into a successful system can be largely facilitated by the use of models, where interaction between the factors can be seen and different options can be tested for evaluation.

MODELLING OF INTERMODAL SYSTEMS

As shown, intermodal transport systems are complex systems that involve a multitude of actors and possible decision making. Implementing changes or starting up new intermodal services involves large costs and brings a risk of failure – and in the worst case, bankruptcy. To reduce the risk of failure, models can be used

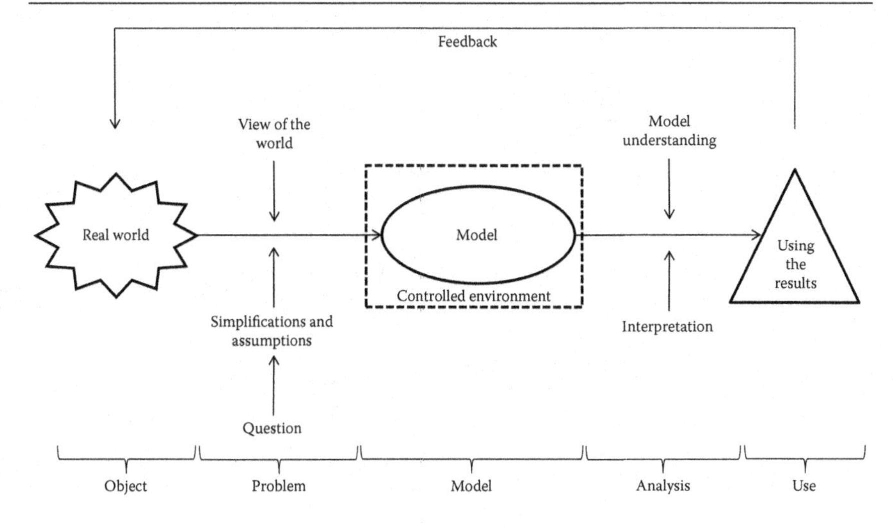

Figure 11.3 The concept of modelling.

to test the system safely via a computer before implementing it in reality. The main purposes for using models include describing, explaining, understanding, predicting and making decisions (Hägg and Wiedersheim-Paul, 1994). A model enables the decision maker to control the studied system and make changes that would not have been possible to make in the real world or that might have been too risky. The overall concept of modelling is shown in Figure 11.3 and is further elaborated in the following sections.

Strategic, tactical and operational models

An intermodal transport system can be analysed on several levels. The system can be divided into three levels: freight flows, transport network and transport infrastructure (Wandel et al., 1992) (see Figure 11.4). The top level represents supply chains in nodes and links and is the demand for transport, for example the number of shipments, size, time constraints, frequency and so on. The second level represents the transport network and its associated traffic, for example the movement of trucks and trains on a transport network. This is derived from the freight flow level. The traffic is shown on the bottom level, which consists of the infrastructure in the transport system. The levels are connected by markets that match the demand and supply of the different levels.

Intermodal transport models can work on any or all three levels and are divided into three main types of models: strategic, tactical and operational models (see Table 11.1). Operational models concern day-to-day operations, such as the scheduling and routing of vehicles. These models focus on commonly occurring and structured problems, meaning that these are routine, repetitive decisions that can be solved using existing standard solutions. For example, there are known

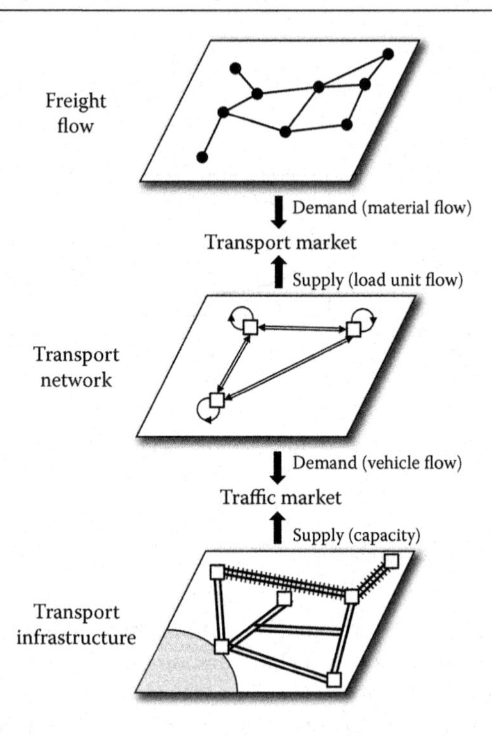

Figure 11.4 Levels of a transport system. (Based on Wandel, S. et al., *Logistiska framsteg – Nordiska forskningsperspektiv på logistik och materialadministration*, Studentlitteratur, Lund, 1992.)

Table 11.1 Model and decision types

Level	*Decision type*	*Decision characteristics*	*Information characteristics*
Strategic	Unstructured	Ad hoc, infrequent, lack of detailed data support	Wide scope, summarised, specialised
Tactical	Semi-structured	Combines structured and unstructured decision making	A mix of data intense, detailed data and wide scope unstructured data
Operational	Structured	Predictable, rational, frequent, narrowly focused	Data intense, detailed

Source: Flodén, J., *Essentials of Information Systems*, Studentlitteratur, Lund, 2013.

procedures and commercially available software that determine how to best calculate the routing of vehicles. The solutions are often rational and fact based with little room for personal opinion. These models often occur on the lower levels of the transport system, such as the traffic market. The models are detailed and data intense, that is they require a lot of detailed input data but are often less complex as models because the problems are well known and structured.

The opposite of the operational model is the strategic model, which concerns long-term strategic decisions, such as should we start an intermodal service on a foreign market or completely change our business model? These are rarely occurring and unstructured problems, meaning that they are unique with no pre-existing solution procedure. There are many unknown factors in the decision, and the available input data is often limited. These models are less detailed but more complex due to the unstructured problem and wide scope with many unknown factors to consider and input data required from many areas. Sometimes, several different models with different scopes are combined to form a strategic model. The solutions are also less detailed and require a lot of interpretation and analysis before being used for decision making. They often occur at the top levels of the transport system, such as the transport market.

Tactical models are a mix between operational and strategic models. These concern semi-structured problems on a medium time frame and contain both structured and unstructured parts. For example: which rail operator should we use for our intermodal service? This is a structured problem where we can calculate and compare prices, but there is also an unstructured portion concerning the trustworthiness, quality, flexibility and so on of the rail operator.

Focus the model

A model is a tool that intends to answer a question; without a question, the model has no purpose. Building and operating a model can be very time and resource consuming and should not be undertaken before a thorough analysis has been made regarding the problem that needs to be solved and it has been determined that a model is the best tool to solve the problem. The model should therefore focus on the question it intends to answer, which means that a model will include simplifications and assumptions that are appropriate to answer the question. Intermodal transport systems are large and complex systems; it is not possible to include all parts in the model, and it is also unnecessary to include parts that will not help answer the question. For example, if road transport is purchased from a sub-contractor at a fixed rate, it is unnecessary to include a detailed modelling of the drivers' working hours, overtime pay, pension fees and so on. An intermodal transport model is, thus, a simplification of reality. It is easy to believe that the more detailed a model is, the better it is, but that is not the case. There is no purpose to having a detailed model just for the sake of having a detailed model, unless this is required by the question the model should answer. Over-parameterisation (using too many variables) and over-elaboration are considered signs of a weak model, while a simple but correct model is considered strong. Adding excessive elaboration to an already weak model will not make it more correct (Box, 1976).

Many simplifications can be derived from the questions to the answers, but others have more practical reasons. The input data required by the model must be available, and the model must also be able to be built and operated in time to provide useful input to the decision making. Although computers are becoming increasingly powerful, it is not difficult to build a model that takes years to solve. For example, solving the travelling salesman problem (a classic problem about visiting all cities in the best order without visiting any city twice) for 24,978 Swedish villages would take 84.8 years on a standard computer (Applegate et al., 2007). Obviously, simplifications must be made; the challenge is to find simplifications that allow the model to answer the questions while making the model practically usable. It is important to recognise that because all models are simplifications of reality, they are also always wrong. The interesting issue is how wrong they are, and if they can be used to answer the problem; as famously stated, 'All models are wrong, some models are useful!' (Box and Draper, 1987). It is very important to be aware of this when using models.

View of the world

Models are depictions of the real world; however, the view of the real world is not independent to the observer. Different persons view and interpret the world differently. A classic example is ambiguous pictures, where two pictures are hidden in the same picture; what you view depends on what you focus on (see Figure 11.5).

In modelling, it is therefore important to realise that the model will also be dependent on the modeller, as two modellers are not likely to build the same model, even if faced with completely identical problems (Arbnor and Bjerke, 1994). To highlight the perception of the intermodal transport system, conceptual models can be drawn. A conceptual model is a graphical representation of the intermodal system that shows key areas, their relationships and borders. Each

Figure 11.5 Different views of the world. (a) Rabbit or duck? (From Fliegende Blätter, Kaninchen und Ente [Rabbit and Duck], 1892). (b) Old woman or young woman? (From Hill, W.E., *Puck*, 78, 11, 1915.)

actor in the intermodal system will have its view of where the focus of the intermodal system lies; at the same time, the modelling itself is often done by external model experts – sometimes with limited experience of intermodal transport. It is therefore important to ensure a common view of the system that is modelled in order to gain acceptance from all actors and help validate the model. Further, the conceptual model highlights the borders and level of detail of the system that is modelled. For example, an intermodal model focused on train operation might be interested in modelling many actors in the rail sectors and the trains in detail while disregarding actors outside the rail system, such as forwarders or governments. In contrast, a model intended to predict a future modal split might want to include government incentives and competing transport modes while only considering rail operations at a basic level. The conceptual model should be validated by the involved actors before the model building starts. An example of a conceptual intermodal model can be seen in Figure 11.6.

Input data

When building an intermodal transport model, the input data availability must also be considered. There is no use spending resources to build a model that cannot be supplied with the appropriate input data, which is why the available input data should be investigated before the model is designed. In a model, the exact format of the input data is important, and the available data should be checked

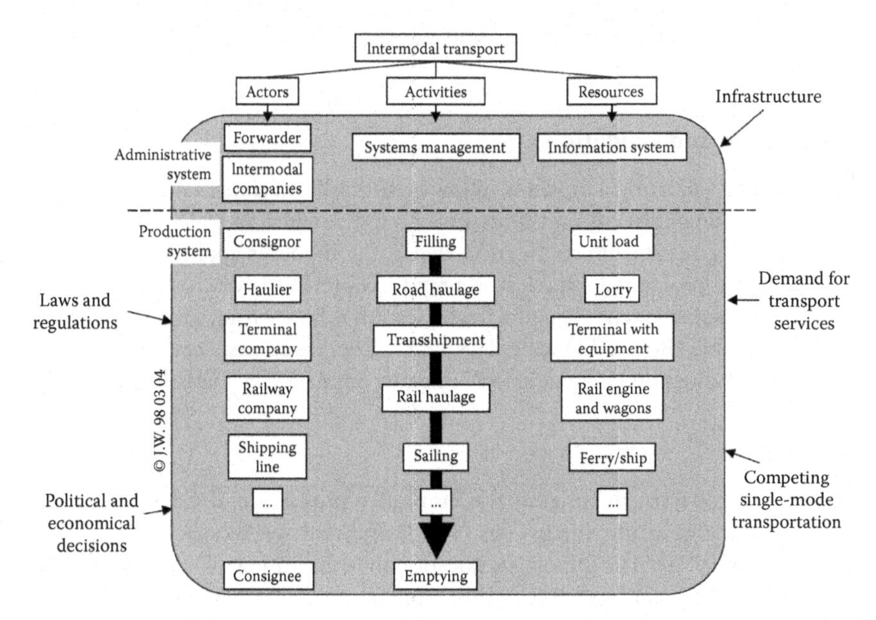

Figure 11.6 A conceptual model of an intermodal transport system. (From Woxenius, J., *Development of small-scale intermodal freight transportation in a systems context.* Doctoral thesis, Chalmers University of Technology, Göteborg, 1998.)

in detail. Even if it is known that, for example transport flow data is available, it is not certain that the format, quality and level of detail in the data are sufficient for or complementary to the model. The data could be partly missing or of low quality, for example it could have inconsistencies, contradictions or missing variables. It is not uncommon for the management of an organisation to truly believe that they have all the data and promise to supply the modeller with detailed data, whereas on later examination, it turns out to be low quality or non-existent. Even if the model is perfect, the model output can never be better than the input data quality, or as jokingly stated, 'Garbage in = garbage out'.

Simplifications

Simplifications in intermodal transport models can take many forms. Some are derived from the question to answer, but others have more practical reasons. Lack of input data or computing capacity often results in a reduced level of detail in the model, for example by aggregating demand to larger groups. Similarly, average costs or a limited number of vehicle types are often used. One of the most difficult aspects to model is human choice, as human decision makers consider a multitude of aspects and are not always fully rational. This is normally managed by the 'economic man' assumption, which states that all actors are rational and fully informed. Examples of common simplifications in intermodal transport models can be found in Table 11.2. Besides the simplifications listed in Table 11.2, the actual optimisation process (i.e. the mathematical calculations) also commonly uses simplifications and heuristics to solve the calculations (see e.g. Nash and Sofer, 1996).

Verification and validation

Verifying and validating a model is an important task and aims to determine if the model is working correctly and gives a true representation of reality. The focus is on researching reasonable confidence, that is the model is applicable to solving the real-world problem at which it is aimed. As previously discussed, all models are wrong to some extent. The practical question is how wrong can they be while still being usable (Box and Draper, 1987)? The process consists of two parts: verification and validation (Lehman, 1977; Pegden et al., 1995; Banks et al., 2001).

Verification

Verification concerns debugging the computer program and establishing that the program runs as intended (Gass, 1983) and that the model has been built right (Banks, 1998). The important question in verification is 'Does the program operate correctly'? and not 'Does this program adequately represent its model and produce output that resembles the real world'? (Lehman, 1977: p. 224). The verification focuses on the internal workings of the model and confirms that the computer program works correctly.

Table 11.2 Examples of simplifications in intermodal transport models

Level	*Example of common simplifications*
Freight flow	Aggregating to fewer origins/destinations. Aggregating commodity types. Using one type of flow measurement (e.g. tonnes, even if volume is the constraining factor for parts of the flow). Assuming all flows can be co-loaded (share the same transport).
Transport market	Not considering competition between transport modes. Rational customers always choosing fastest/cheapest transport options and disregarding other factors, such as reliability. Using a limited number of transport options. Assuming all flows between A and B use the same transport chain.
Transport network	Using average transport cost per tonne (not considering the cost of individual vehicles). Using only direct transport A–E; not 'milk runs' A–B–C–D–E. Not considering the return flow of empty vehicles and load units. Simplifying transhipment processes. Disregarding delivery time requirements. Not considering specific customer requirements (e.g. requirements on load units, delivery, etc.).
Traffic market	Limiting the number of vehicle types available. Not considering competition between transport actors. Assuming vehicles are always available at a given price. Assuming there is continuous and/or unlimited transport capacity (not considering the departure of individual vehicles). Not considering working time regulations for drivers.
Transport infrastructure	Using direct linear distance instead of actual (network) distance. Using average speeds. Not considering delays and capacity constraints. Not including all road and rail lines.

Validation

Validation aims to research reasonable confidence so that the model is applicable to solving the real-world problem at which it is aimed. Validation is focused on the relationship between the model and the outside world. It tests the agreement between the model and the real world being modelled and considers the assumptions and generalisations made in the model. It should confirm that we have built the right model, as opposed to the previous verification step that checks if the model has been built right (Banks, 1998). However, validation is not a simple task. Normally, a comparison can be made between known data from the real-world system and the output of the model. Unfortunately, this is not possible for most intermodal transport models because they include many simplifications and focus on non-existing systems or future events, such as how a new intermodal system should be designed. Further, the models, particularly the strategic and tactical models, are placed in a business environment with independent customers, competitors and so on. This is a societal system whose behaviour cannot be perfectly predicted. It is, for example, impossible to know if your main customer will go bankrupt next year, or if the government will decide to change the tax levels. In a deterministic (i.e. predictable) system, such as physics or chemistry, it is possible to perfectly predict (given the right model) the outcome of a certain experiment. However, in a societal system, this is not possible, as it is simply impossible to perfectly predict the future. The real world being modelled is, therefore, always unknown to some extent. Most intermodal transport models can thus never be completely validated, but only invalidated (Quade, 1980). Therefore, the focus on the validation process must be on trying to invalidate the model, that is to prove it wrong. Quade (1980: p. 34) states: 'When you have tried all the reasonable invalidation procedures you can think of, you will not have a valid model.... You will, however, have a good understanding of the strength and weaknesses of the model....Knowing the limits of the model's predictive capabilities will enable you to express proper confidence in the results obtained from it'. In the theory of science, the principle of invalidation is well known and best explained by Popper (2002): If your hypothesis is that all swans in the world are white, it is better to try to find one black swan than to determine the colour of all swans in the world.

Using the results

Model results cannot just be accepted at face value; they must be interpreted in relation to the scope, assumptions and simplifications made in the model. If a model presents results with many decimals, it does not mean that they are valid to the last decimal. This is particularly important to remember as a decision maker receiving output from someone else's model. Potential error sources include inappropriate simplifications and assumptions, technical faults in the model (e.g. programming errors) and faults in the input data. In general, intermodal transport models are used as decision support systems (DSS) to help support decisions. Due to the complexity of intermodal transport, the models are rarely used to directly manage the system, but rather as support tools where they combine with other data sources; the decision maker's experience makes up the foundation for decision making. Care should be taken to not have an over-reliance on the model or overanalyse the results. For example, traffic forecasts used to determine investments in large infrastructure projects have been shown to be very susceptible to both uncertain forecasts and 'wishful thinking' when interpreting the results, thus overestimating the impacts of the investment (Flyvbjerg et al., 2002, 2005, 2006).

Sensitivity analyses are often used to try to capture uncertainties. This means that the model is run with different input data several times to test the sensitivity of the assumptions made. For example, if future demand is uncertain, the model is run with different demand input to determine how the output is affected. Often, some of the input variables have very little impact on the end result, while only small changes in other variables can have a very large impact. The aim of the sensitivity analysis is to identify how sensitive the end results are to changes in uncertain variables (Hägg and Wiedersheim-Paul, 1994).

FOUR-STEP MODEL

Traditionally, freight transport modelling is divided into four steps according to the classical four-step model (see Figure 11.7). The four-step model is not a 'model' in itself, but rather a methodology of steps that are performed as part of a complete transport model. The four-step model was originally designed for passenger transport, but is also commonly used for freight transport (McNally, 2000).

1. Production and attraction
2. Trip distribution
3. Modal split
4. Assignment

Note that a transport model does not have to include all four steps, although some models do. Similarly, it is possible to iterate between the steps, for example when the selected modal split (Stage 3) and its characteristics are allowed to influence the trip distribution (Stage 2) (Bates, 2000; McNally, 2000). A selected mode might have properties (e.g. speed) that contrast with the requirements used for trip distribution. The trip distribution can then be performed again using the new input from the modal split. The four steps will be explained in the following sections.

Production and attraction

This first step determines the total goods volume going in and out of a region. This step is not concerned with where the goods are going to or coming from – only with the production and attraction of a region. When proper freight statistics are missing, this is normally performed by econometrics and statistical calculations based on geographical characteristics, such as industry structure, company turnover, population or gross domestic product (GDP). A review by

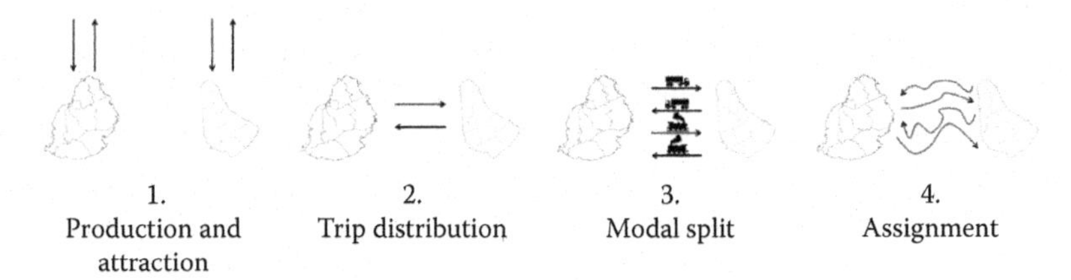

Figure 11.7 The four-step model.

ME&P and WSP (2002) defines three main approaches to production and attraction in transport models:

1. Trend and time series models: The extrapolation of historical data into the future, for example regression models or time series.
2. System dynamic models: System dynamics is a mathematical modelling technique that studies the behaviour of complex systems over time. Transport demand develops and changes over time using feedback loops to/from land use, economy and so on.
3. Input-output models: Macroeconomic models where demand for transport is derived from economic activity. This uses input-output matrices to show how the output of one industry is an input to other industries, for example to produce one tonne of product type X, one tonne of product type Z and one tonne of product type Y are needed.

Trip distribution

The second step concerns how the demand for transport is translated into origin/destination (O/D) pairs, that is which good is transported from where to where. This is based on the production and attraction of a region from Step 1 and uses factors such as transport costs and distances. The review by ME&P and WSP (2002) defines two approaches to trip distribution. The most common is the gravity model, where an interaction between two locations is assumed to decline with an increasing distance between them due to increasing transport cost and time. The other main approach is to use regional input-output models, that is performing production and attraction (Step 1) with regional input-output matrices, thus receiving the trip distribution as an output from that step. This means that the input-output matrix, which shows the input/output from a type of industry, also shows the region the input comes from or goes to, for example 80% of all steel comes from region X. This can be made by, for example combining an input-output model with trade statistics.

Modal split

The modal split step concerns the choice of transport mode, for example if the freight should be transported by road or rail. This selection is, in reality, most often dependent on a number of factors, such as transport cost, time, quality, infrastructure and so on (Flodén et al., 2017); however, the decision for simplicity is made by cost in many models, where the transport mode that offers the lowest cost is selected. It is possible to separate between aggregated and disaggregated modal split models (ME&P and WSP, 2002). Aggregate models work on a zone level and determine the share for each transport mode per zone, where disaggregate models determine the modal split for each shipment.

A common method for the modal split is the logit model. Logit models are based on a random utility function that shows how much a certain alternative is worth to a decision maker. The logit model consists of a number of exhaustive and mutually exclusive options, that is the decision maker must choose only one of the options. Each option has a given utility value, which is a number representing how valuable the option is for the decision maker. This is based on the observable attributes of the option, for example speed and cost, and the observable characteristics of the decision maker, for example type of industry. The decision maker is assumed to choose the option with the highest utility value. However, in reality, it is impossible to perfectly predict the decision maker's choice as it is impossible to perfectly measure the preferences of the decision maker and all possible attributes of the options. These unobservable components are included as a random component in the model to represent the measurement errors. Mathematically, this can be expressed as in Equation 11.1, where *Ui*,q is the utility of an alternative i to an individual q. The utility consists of a deterministic component V and a random component ε. Based on this, the likelihood that a certain option is selected can be calculated (Koppelman and Sethi, 2000).

$$U_{i,q} = V_{i,q} + \varepsilon_{i,q} \tag{11.1}$$

A multinomial logit model makes a choice between a number of equal options. However, sometimes a more realistic representation of the modal choice is to make the choice in steps, for example a traveller might make a choice between using a car, bike or public transportation, but the type of public transport used is only selected as a second step after the decision to use public transport has been made. A logit model that allows for this is called a *nested logit model*, where choices with the same attributes are put in a 'nest' (see Figure 11.8, where the logit models are depicted as a tree structure).

Multimodal network models can also be used when the network (including all transport modes) is defined as a number of links and nodes (e.g. terminals and transhipment points) with certain attributes, for example cost and time to use them. A cost-minimising algorithm is used to determine the modal split and route choice in the network. Figure 11.9 is a representation of a real network (a), with nodes Xa, Xb, Xc and Xd, and a network model representation (b) of the same network (TERMINET, 2000: pp. 93–94). Note that extra links have been added where several modes are available (Xa to Xb), and all possible transhipments are represented by links to include the modelling of intermodal transport. Using each link incurs a cost; a cost-minimising algorithm can be used to find the lowest cost path through the network.

The modal split might also be externally given to the model, that is as part of the input data.

Assignment

This last step concerns the route choice in the network, for example which roads to use between A and B. The assignment is often based on selecting the shortest route but also other criteria could be used, for example selecting the cheapest route. In most cases, the cheapest and shortest routes are the same as many costs are distance

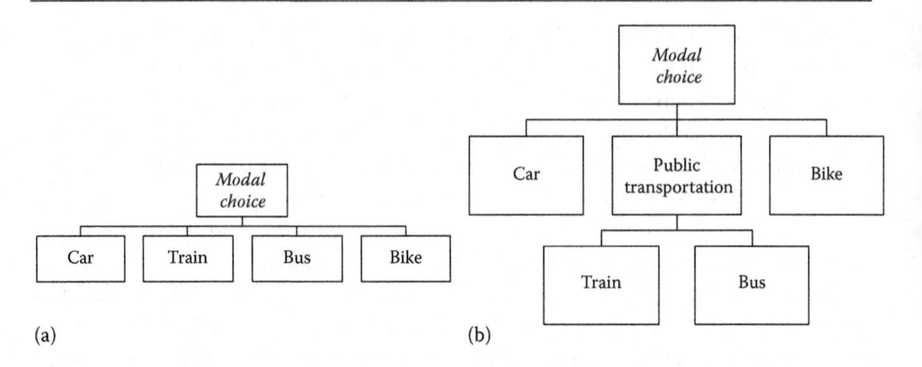

Figure 11.8 Logit models. (a) Multinomial logit model. (b) Nested logit model.

dependent, although this might depend on for example road tolls and transhipment costs. The route selection can also be subject to constraints, such as selecting the cheapest route that meets the required delivery times (see Figure 11.10).

Shortest-path algorithms can be used to find the preferred route through the network. The assignment step can also be included in the modal split step when a multimodal network model is used.

AGENT-BASED MODELS

Since the 1990s, agent-based models (ABMs) have found their way into the field of freight transport modelling (Tavasszy, 2006). As the name suggests, ABMs model the behaviour of actors in complex systems. The overall system performance is then the result of all actors' interactions. Whereas four-step models are generally fast and relatively straightforward, ABMs are more complex and demand higher computational efforts (Mommens et al., 2016). The important advantages of ABMs include their ability to better represent transport flows because of their disaggregated nature and incorporating agent behaviour. Both should lead to better interpretation and evaluation of the model results.

In ABMs, agents can make autonomous decisions based on the (limited) information made available to them and try to meet their preferences as well as possible. The ability of actors to interact is paramount to the decisions they can and will make. The modeller thus needs to define the information availability among agents (groups) and their capacity to influence other decision makers. The agents in a model will thus have incomplete information about the context and limited influence on other agents' decisions (Reis, 2014).

Figure 11.11 presents a fictitious example of different transport agents' interactions in an ABM. It is clear that no agent has access to all available information. As a consequence, the model results satisfy the needs of the agents with sufficient bargaining power, but they will most probably not present an optimal solution for the whole system. An existing example of an intermodal ABM is the transportation and production agent–based simulator (TAPAS) (Holmgren et al., 2012).

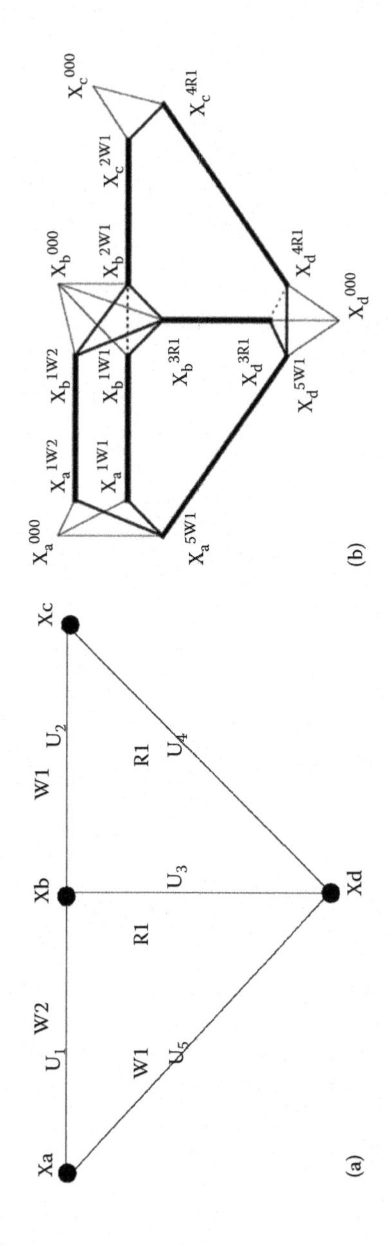

Figure 11.9 Network models. (From TERMINET, Towards a new generation of networks and terminals for multimodal freight transport. Final report. EU 4th framework research programme, 93–94. Technische Universiteit Delft, 2000.)

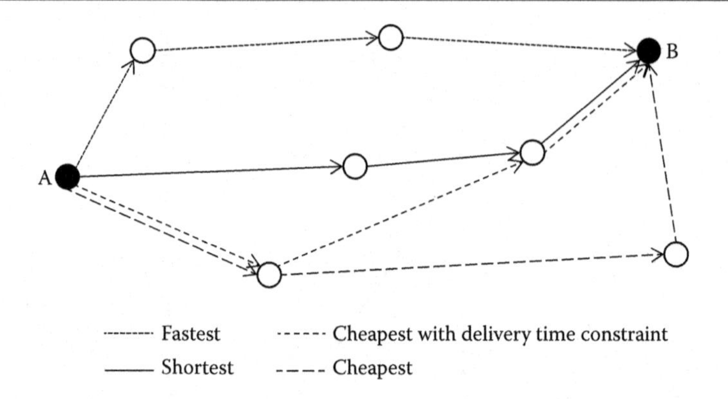

Figure 11.10 Different assignment options from A to B.

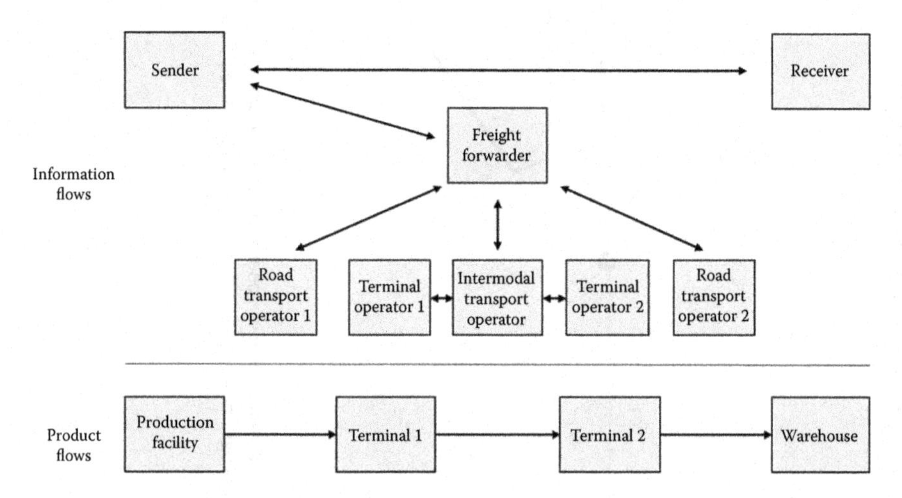

Figure 11.11 Example of agents' interaction opportunities in an intermodal ABM. (Based on Reis, V., *Transportation Research Part A: Policy and Practice*, 61, 100–120, 2014.)

This attempts to model the behaviour of customers, transport chain coordinators, product buyers, transport buyers, production planners and transport planners. As in real life, not all actor groups interact. This model can simulate the different transport agents' reactions on the implementation of policies.

LAMBIT MODELLING SYSTEM

Intermodal modelling systems exist in many different sizes. The simplest ones are just Excel spreadsheets, while other models are advanced, stand-alone modelling systems. This chapter will give an example of one extensive stand-alone modelling system, the geographic information systems (GIS)-based location analysis model for Belgian intermodal terminals (LAMBIT). The main goal of this model is to provide evaluations of policies that could possibly affect the use of intermodal transport.

Some intermodal freight transport models use GIS extensively. This section briefly describes the functionalities of such models and their possible applications. According to Shaw and Rodrigue (2016), GIS are information systems that specialise in the input, management, analysis, display and reporting of geographical (spatial) information. The most obvious advantage of GIS-based models is that they allow one to easily visualise model input and output, which can help determine spatial relations. Intermodal transport planning problems can be subdivided into strategic, tactical and operational models (see Table 11.1). The focus of GIS-based models is mainly the interface of strategic and tactical decisions concerning the infrastructure network evaluation. Two distinct types of network analyses can be performed: the first focuses on altered choice behaviour due to changes in transport costs as an effect of policy measures or the use of different vehicles; it can be categorised under network evaluation models. The second type of analysis focuses on the addition of new or the elimination of existing transport infrastructure.

Existing GIS-based models for intermodal network evaluation mainly take the perspective of policymakers and transport or network operators. Policymakers can try to stimulate the use of intermodal transport using different measures, such as taxes, financial incentives and infrastructural changes. From the perspective of transport and terminal operators, the effects of using new or altered vehicles can be modelled. A clear advantage of GIS is its ability to spatially represent transport networks, origins, destinations and transhipment locations. Using GIS in intermodal network models provides an interface in which to evaluate the effects of measures that aim to improve mode-specific transport services (Boile, 2000). LAMBIT is a network evaluation model that can estimate the impact of changes on transport costs, external costs, modal splits, service areas and so on, as opposed to network optimisation models that try to design networks or services in order to optimise parameters. The model is able to deal with the previously described types of evaluations, namely, simulating the effects of cost and network changes.

GIS-based analysis methodology

The main goal of the LAMBIT methodology is to test the impact of different scenarios concerning the competitiveness of intermodal transport solutions. It is an evaluation methodology that has been used for analyses of the Belgian intermodal landscape. However, the methodology and possible applications discussed can easily be transferred to other case studies when the necessary input variables are available.

In this section, the methodology is discussed by describing the different building blocks of the model. The LAMBIT methodology entails three major components (Figure 11.12). The first component consists of the model inputs, which are divided into three distinct input data categories. The core component of the model is a GIS-based intermodal landscape simulation model, which can be used for a range of scenario analyses. The output of the model can be of different kinds depending on the input data and the problem analysed. The different subcomponents of the methodology are discussed next.

Input parameters

LAMBIT uses three types of input parameters for simulations of intermodal transport systems. The reliability of the results of LAMBIT analyses obviously depends on the reliability of the input data.

The first input for the LAMBIT analysis is the transport demand in terms of the goods flows in the studied region. The format of this O/D matrix depends on the research question that is tackled. LAMBIT has mainly been used to look into maritime-based container transport (e.g. Macharis and Pekin, 2009).

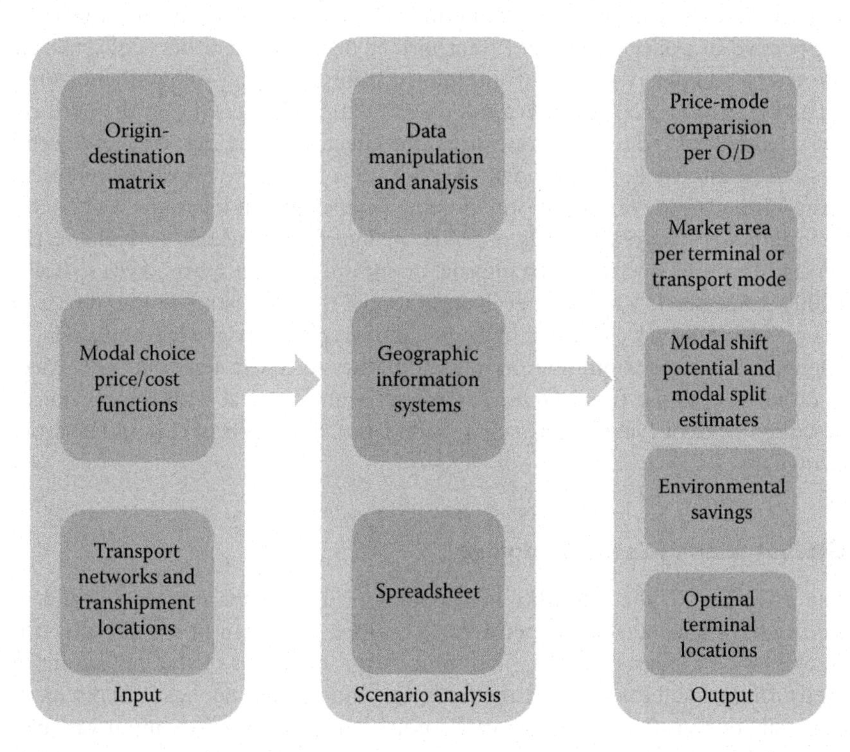

Figure 11.12 Architecture of the LAMBIT model. (Based on Pekin, E., *Intermodal Transport Policy: A GIS-based Intermodal Transport Policy Evaluation Model*, Vrije Universiteit, Brussel, 2010.)

The transport network layers are a second core component of the model. In the GIS software used for the analysis,* these networks can be represented by vector files representing real-world transport networks. The transport network layers might include maritime, road, rail, inland waterway, air and short sea shipping networks. These are used for routing, necessitating transhipment locations as interchanges between transport modes, such as seaports and inland terminals. Finally, the locations of origins and destinations are included for the purpose of routing. These locations might be addresses, or when general statistics are used, they can be centres of corresponding geographic entities. In the analysis stage, the combination of these networks and point locations allows for the calculation of possible routes between any origin and destination using the available transport modes or a combination of modes.

The third input is unimodal and intermodal cost functions. A comparison between the intermodal cost structure and the unimodal road transport allows for assessing the competitiveness of these transport chains. Basically, the total cost of an intermodal transport service is composed of the costs of main haulage, drayage and transhipments (see Chapter 8). The advantage of intermodal transport, compared with unimodal road transport, lies in the smaller distance-dependent variable costs during main haulage, which are the result of the scale economies that are obtained by the large volumes that can be transported in a single transport movement. Each combination of transport modes and each single mode have their own cost functions, allowing for the calculation of the total cost of each possible route. Next to direct costs, modal choice decisions can also be influenced by qualitative criteria measuring the performance of transport chains. These factors can also be included in the cost functions to provide total logistics cost functions. These criteria should, however, be translated into cost components. Examples of such valuations are the values of transport time, transport time reliability, frequency, order time and environmental criteria. The parameters in these (total logistics) cost functions can be derived from the literature (e.g. Janic, 2007) or field surveys.

Model operation

Different algorithms can be used to calculate relevant routes in GIS software. Possibilities include (variants on) the shortest-path algorithm developed by Dijkstra (1959), which minimises the modal choice cost functions for each transport alternative. The LAMBIT methodology explores the relative attractiveness of all transportation alternatives through price or cost minimisation. Therefore, relevant price or cost variables are linked to the corresponding network layers and transhipment locations, including the origin and destination of the transport.

* Until now, the model has been run on ArcGIS software, but other GIS software, such as QGIS, could be used.

The price or cost to travel a link or node is attached to each network segment and transhipment location. For each O/D combination, all possible calculated routes can be compared by their (total logistics) cost performance or environmental criteria.

Output

The five output types mentioned in Figure 11.12 can be derived from this minimisation. The most basic outputs are the cost comparisons for the different transport modes per O/D couple. This information can be converted into the (spatial) market area of transport modes, as shown in the next section. The modal shift potential can be estimated from the market area analysis by linking these market areas to a freight transport data matrix. The potential for a modal shift using a specific terminal can be derived from the volume currently transported by road transport to the locations within this terminal's market area.

One advantage of this GIS-based methodology is that the input and output can be used in other analyses, and the model can also be used in complementarity with other models, as has been shown in Macharis et al. (2008), where the model was linked to NODUS, a network model, and SIMBA, a discrete event simulation model. However, the main goal of LAMBIT is to simulate the impact of changes in the input parameters of the model. It is interesting how the previously described output parameters react to such changes. The next section will therefore focus on the possibilities for applying decision support analysis using a GIS-based analysis methodology.

Scenario analyses

To stimulate intermodal transport, several policy measures are available. A reference scenario with the existing intermodal connections, current infrastructure network and current market prices serves as a benchmark. LAMBIT can assess scenarios impacting (a combination of) the input parameters (Table 11.3). As previously stated, the actual output of the analysis depends on the problem statement, but it can be found within the five output categories mentioned in Figure 11.12.

The output from a reference scenario is shown in Figure 11.13a, which depicts the existing intermodal inland terminals in Belgium with their corresponding market areas and considers transport to and from the Port of Antwerp. This reference scenario is based on current intermodal services to the Port of Antwerp. Municipalities are highlighted when intermodal transport has a more attractive transport price compared with unimodal road transport. Two examples of scenario evaluations are briefly described.

In Belgium, subsidies were granted for the transhipment of containers at intermodal terminals in certain cases. The effect of a rail transport subsidy on the extent of intermodal terminal market areas was simulated by Macharis and

Table 11.3 Examples of LAMBIT scenarios

Input category	*Possible scenario*
O/D matrix	Extending spatial extent database
	Individual company analysis
Modal choice cost function	Intermodal transport/transhipment subsidies
	Internalisation of external transport costs
	Infrastructure taxes or pricing
	Impact of additional modal choice criteria (e.g. transport time)
	Vehicle innovations (e.g. longer and heavier vehicles)
Infrastructure	Addition/reduction of intermodal terminals
	Intermodal network changes (e.g. new rail line)

Pekin (2009) and updated by Macharis et al. (2010) to account for barge subsidies. The subsidy schemes were not constant over time, not equal in space and not equal for each transport mode. The latter analysis shows that clear competition between intermodal barge and intermodal rail transport was witnessed when only one mode was subsidised, and when both subsidy schemes were implemented simultaneously, most intermodal terminal market areas increased in size. However, depending on the relative location of each transhipment terminal, some barge terminals decreased in market area extent because of the increased competition from intermodal rail transport and the mutual competition among barge terminals (Figure 11.13b). Uncoordinated subsidy schemes might thus induce a modal shift from barge to rail transport, whereas this was not the original goal of the subsidy schemes.

A second example relates to changes in the intermodal terminal landscape following the introduction of new or the closing of old inland terminals. Other examples of infrastructural changes are the construction of new roads and railways or the widening and deepening of canals, which impact the types of vehicles that can be used on the corresponding network segments. Additional terminals can increase accessibility to intermodal transport services, but an abundance of new terminals might harm the competitiveness of individual terminals. In order to evaluate the potential of (potential) new terminals, LAMBIT can be used to analyse the market area and potential transhipment volume of these terminals.

In Figure 11.14, nine potential new barge terminals and two rail terminals are added to the current terminal landscape. The future terminal landscape indicates growth for the market areas of barge terminals. Overall, barge terminals in the south of the country can take market areas from the unimodal road transport. However, implementing all the proposed terminals at once will harm the intermodal transport sector as increased competition between terminals arises. To quantify the extent of a shift between terminals, detailed information on transport volumes should be available.

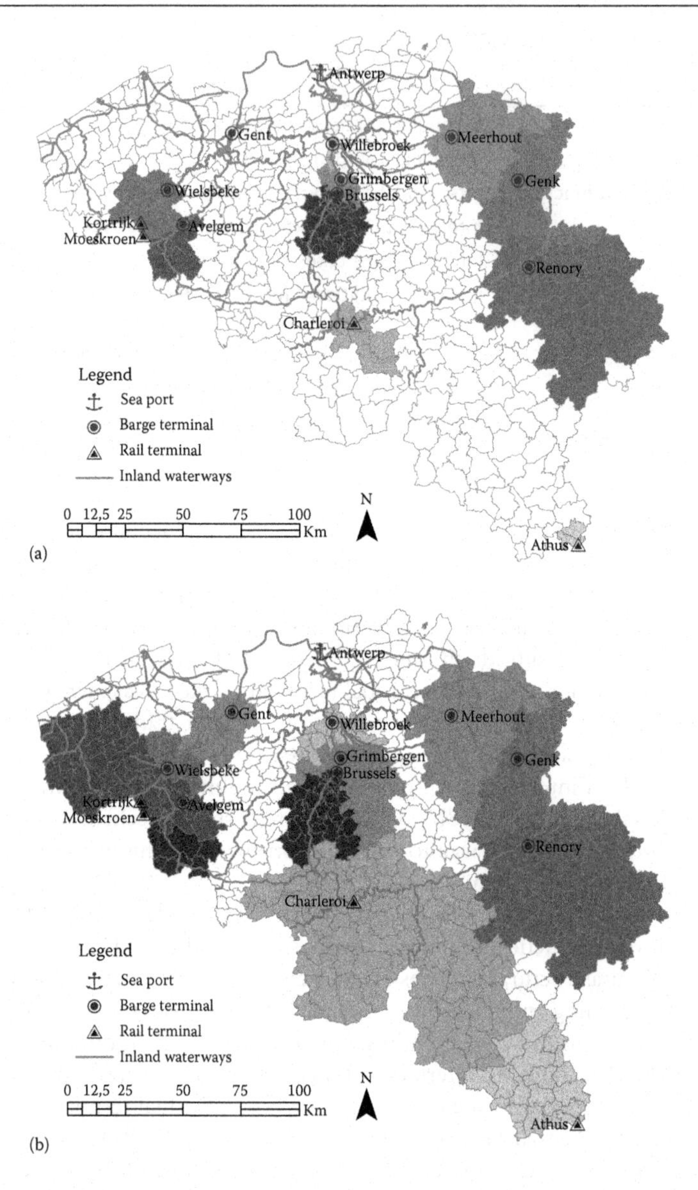

Figure 11.13 The effects of subsidy scenarios for inland waterways and rail transport anno 2008 (b) compared with the reference situation (a). The subsidies bring increased potential for intermodal transport, but competition among terminals might occur. (Based on Macharis, C. et al., The Internalisation of External Cost: Opportunity or Threat for Intermodal Transport? Paper presented at NECTAR Conference and published in proceedings, Washington, 2009; Macharis, C. et al., Intermodaal Binnenvaartvervoer: Economische en ecologische aspecten van het intermodaal binnenvaartvervoer in Vlaanderen, VUBPRESS, Brussel, 2011.)

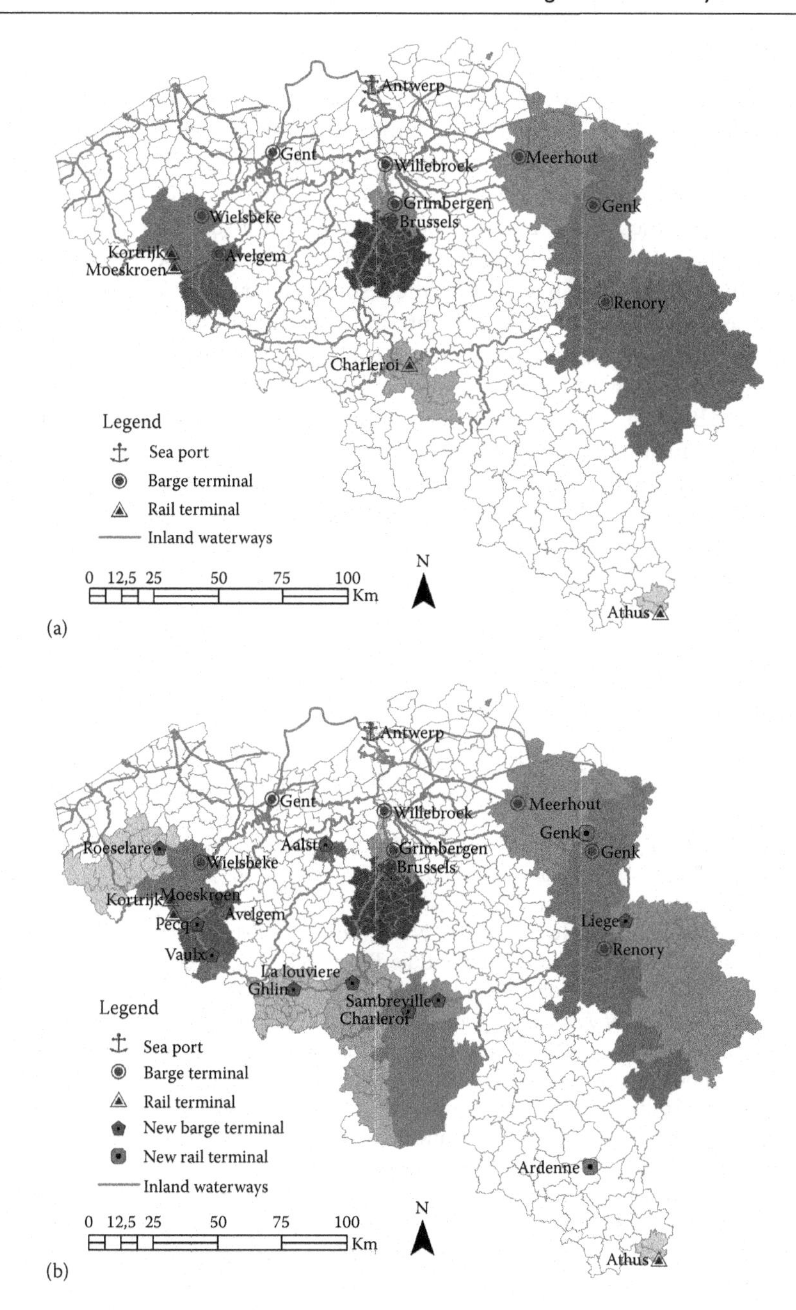

Figure 11.14 The intermodal terminal market areas when 11 new terminals are added (b). (Based on Macharis, C. et al., Intermodaal Binnenvaartvervoer: Economische en ecologische aspecten van het intermodaal binnenvaartvervoer in Vlaanderen, VUBPRESS, Brussel, 2011.)

CONCLUSION

Intermodal transport systems are more complex than unimodal transport systems, which makes it harder to assess the impact of decisions on system and service designs. Decision making can be supported by the use of models. Models are tools intended to help answer a question and support decision making. The complexity of the system makes simplifications and assumptions necessary in the modelling. It is therefore important to understand the model as well as the assumptions made when interpreting the results. The traditional four-step modelling methodology can be used for intermodal models. This model consists of four main steps: (1) production and attraction, (2) trip distribution, (3) modal split and (4) assignment, although many models do not include all steps. More recently, ABMs have found their way into intermodal modelling, which allows for better simulating actor behaviour.

The straightforward LAMBIT methodology is described as an example of a decision support model for policy measures and network analyses concerning the competitiveness of intermodal transport. The proposed methodology facilitates easy inclusion of all relevant transport modes and transhipment locations for a large-scale study area.

REFERENCES

Applegate, D. L., Bixby, R. E., Chvatal, V., Cook, W. J. (2007). The Traveling Salesman Problem: A Computational Study. Princeton University Press. Princeton, NJ.

Arbnor, I., Bjerke, B. (1994). *Företagsekonomisk metodlära*. Studentlitteratur. Lund.

Banks, J. (1998). *Handbook of Simulation : Principles, Methodology, Advances, Applications, and Practice*. Wiley. New York.

Banks, J., Carson, J. S., Nelson, B. L., Nicol, D. M. (2001). Discrete-Event System Simulation. Prentice-Hall. Upper Saddle River, NJ.

Bates, J. (2000). History of demand modelling. In: Hensher, D. A., Button, K. J. (eds.) *Handbook of Transport Modelling*, pp. 11–34. Pergamon. Amsterdam.

Behrends, S., Flodén, J. (2012). The effect of transhipment costs on the performance of intermodal line-trains. *Logistics Research*. 4 (3): 127–136.

Boile, M. P. (2000). Intermodal transportation network analysis: A GIS application. Paper presented at 10th Mediterranean Electrotechnical Conference, MEleCon 2000, Vol. II and published in proceedings 2000.

Box, G. E. P. (1976). Science and statistics. *Journal of the American Statistical Association*. 71 (356): 791–799.

Box, G. E. P., Draper, N. R. (1987). *Empirical Model-Building and Response Surfaces*. Wiley. New York.

Dijkstra, E. W. (1959). A note on two problems in connexion with graphs. *Numerische Mathematik*. 1: 269–271.

Fliegende Blätter (1892). Kaninchen und Ente (Rabbit and Duck). Fliegende Blätter, p. 17.

Flodén, J. (2013). *Essentials of Information Systems*. Studentlitteratur. Lund.

Flodén, J., Bärthel, F., Sorkina, E. (2017). Transport buyers choice of transport service—A literature review of empirical results. *Research in Transportation Business & Management*, Volume 23.

Flyvbjerg, B., Holm, M. S., Buhl, S. (2002). Underestimating costs in public works projects: Error or lie? *Journal of the American Planning Association*. 68 (3): 279–295.

Flyvbjerg, B., Skamris Holm, M. K., Buhl, S. L. (2005). How (in)accurate are demand forecasts in public works projects?: The case of transportation. *Journal of the American Planning Association*. 71 (2): 131–146.

Flyvbjerg, B., Skamris Holm, M. K., Buhl, S. L. (2006). Inaccuracy in traffic forecasts. *Transport Reviews*. 26 (1): 1–24.

Gass, S. I. (1983). Decision-aiding models: Validation, assessment, and related issues for policy analysis. *Operations Research*. 31 (4): 603–631.

Hägg, I., Wiedersheim-Paul, F. (1994). *Modeller som redskap: att hantera fö retagsekonomiska problem* (Models as tools). Liber-Hermods. Malmö, Sweden.

Hill, W. E. (1915). My wife and my mother-in-law. *Puck*. 78 (2018): 11.

Holmgren, J., Davidsson, P., Persson, J. A., Ramstedt, L. (2012). TAPAS: A multi-agent-based model for simulation of transport chains. *Simulation Modelling Practice and Theory*. 23 (April): 1–18.

Janic, M. (2007). Modelling the full costs of an intermodal and road freight transport network. *Transportation Research Part D: Transport and Environment*. 12 (1): 33–44.

Koppelman, F. S., Sethi, V. (2000). Closed-form discrete-choice models. In: Hensher, D. A., Button, K. J. (eds.) *Handbook of Transport Modelling*, pp. 211–228. Pergamon. Amsterdam.

Lehman, R. S. (1977). Computer Simulation and Modeling: An Introduction. Erlbaum. Hillsdale, NJ.

Macharis, C., Pekin, E. (2009). Assessing policy measures for the stimulation of intermodal transport: A GIS-based policy analysis. *Journal of Transport Geography*. 17 (6): 500–508.

Macharis, C., Pekin, E., Caris, A., Jourquin, B. (2008). *A Decision Support System for Intermodal Transport Policy*. VUBPRESS. Brussel.

Macharis, C., Pekin, E., van Lier, T. (2009). The internalisation of external cost: Opportunity or threat for intermodal transport?. Paper presented at NECTAR Conference and published in proceedings, 18–20 June 2009, Washington.

Macharis, C., Pekin, E., van Lier, T. (2010). A decision analysis framework for intermodal transport: Evaluating different policy measures to stimulate the market. In: Givoni, M., Banister, D. (eds.) Integrated Transport: From Policy to Practice, pp. 223–240. Routledge. Abingdon, UK.

Macharis, C., van Lier, T., Pekin, E., Verbeke, A. (2011). Intermodaal Binnenvaartvervoer: Economische en ecologische aspecten van het intermodaal binnenvaartvervoer in Vlaanderen, 9789054878681, VUBPRESS. Brussel.

McNally, M. G. (2000). The four-step model. In: Hensher, D. A., Button, K. J. (eds.) *Handbook of Transport Modelling*, pp. 35–52. Pergamon. Amsterdam.

ME&P and WSP. (2002). Review of models in continental Europe and elsewhere, Report B2, Review of Freight Modelling, ME&P – WSP. Cambridge.

Meers, D., Vermeiren, T., Macharis, C. (2014) Intermodal break-even distances: A fetish of 300 kilometres?. In: Macharis, C., Melo, S., Woxenius, J., van Lier, T. (eds.) *Sustainable Logistics*, pp. 217–243. Emerald. Bingley.

Mommens, K., van Lier, T., Macharis, C. (2016). Loading unit in freight transport modelling. *Procedia Computer Science.* 83: 921–927.
Nash, S. G., Sofer, A. (1996). *Linear and Nonlinear Programming.* McGraw-Hill. New York.
Niérat, P. (1997). Market area of rail-truck terminals: Pertinence of the spatial theory. *Transportation Research Part A: Policy and Practice.* 31 (2): 109–127.
Pegden, C. D., Sadowski, R. P., Shannon, R. E. (1995). *Introduction to Simulation Using SIMAN.* McGraw-Hill. New York.
Pekin, E. (2010). *Intermodal Transport Policy: A GIS-based Intermodal Transport Policy Evaluation Model.* Vrije Universiteit. Brussel.
Popper, K. (2002). *The Logic of Scientific Discovery.* Routledge. London.
Quade, E. S. (1980). Pitfalls in formulation and modeling. In: Majone, G., Quade, E. S. (eds.) Pitfalls of Analysis, pp. 23–43. Wiley. Chichester, UK.
Reis, V. (2014). Analysis of mode choice variables in short-distance intermodal freight transport using an agent-based model. *Transportation Research Part A: Policy and Practice*, 61 (March): 100–120.
Shaw, S.-L., Rodrigue, J.-P. (2016). Geographic Information Systems for Transportation (GIS-T). Routledge. New York.
Tavasszy, L. A. (2006). Freight modelling: An overview of international experiences. Paper presented at TRB Conference on Freight Demand Modelling: Tools for Public Sector Decision Making and published in proceedings. September 25–27, 2006. Washington, DC.
TERMINET. (2000). Towards a new generation of networks and terminals for multimodal freight transport. Final report. EU 4th framework research programme. Technische Universiteit Delft. Delft.
Wandel, S., Ruijgrok, C. J., Nemoto, T. (1992). Relationships among shifts in logistics, transport, traffic and informatics. In: Huge, M., Storhagen, N. G. (eds.) *Logistiska framsteg – Nordiska forskningsperspektiv på logistik och materialadministration*, pp. 96–136. Studentlitteratur. Lund.
Woxenius, J. (1998). Development of small-scale intermodal freight transportation in a systems context. Doctoral thesis, Chalmers University of Technology, Göteborg.
Woxenius, J., Bärthel, F. (2008). Intermodal road-rail transport in the European Union. In: Konings, R., Priemus, H., Nijkamp, P. (eds.) The Future of Intermodal Freight Transport: Operations, Design and Implementation, pp. 13–33. Edward Elgar Publishing. Cheltenham, UK.

Chapter 12

Operations research and intermodal transport

Teodor Gabriel Crainic and Mike Hewitt

INTERMODAL FREIGHT TRANSPORTATION

Intermodal freight transportation occurs in an environment that is shaped by multiple stakeholders: shippers, carriers and government institutions. Shippers, including goods manufacturers, wholesalers and retailers, create demand for transportation services in the form of requests for the transportation of freight between two points. Such requests need not come from a shipper directly, but instead through a logistics service provider. On the other hand, carriers supply the transportation necessary to move freight. Supporting carriers are a host of other companies, such as firms that operate and maintain intermodal terminals, energy companies that provide fuel for vehicles and vehicle manufacturers. Finally, supply is matched with demand in a regulatory environment that is defined by government institutions. In addition, government institutions provide or influence the physical infrastructure that carriers rely on to provide transportation services.

Each of these stakeholders has a different perspective on intermodal freight transportation and, consequently, a different set of key performance indicators (KPIs). For example, shippers are typically focused on how much the transportation of their goods costs as this impacts their profitability. That said, they also care about customer service metrics, including how quickly the goods can be delivered to their customers. Similarly, with the advent of manufacturing policies such as just-in-time manufacturing, shippers that participate in supply chains are increasingly focusing on reliability. Carriers, on the other hand, are typically focused on the costs they incur when offering transportation services. However, as predictors of those costs, they also focus on performance metrics such as maintenance frequency, employee turnover and resource utilisation. That said, carriers are increasingly focusing on sustainability-related metrics, such as emissions and fuel efficiency. Finally, government institutions are focused on the many ways in which freight transportation impacts their constituencies, from how the noise and air pollution emitted by transportation affects their environment, to how the presence or absence of freight transportation can impact their economy. Figure 12.1 illustrates these stakeholders, the issues that drive their behaviour and what they offer.

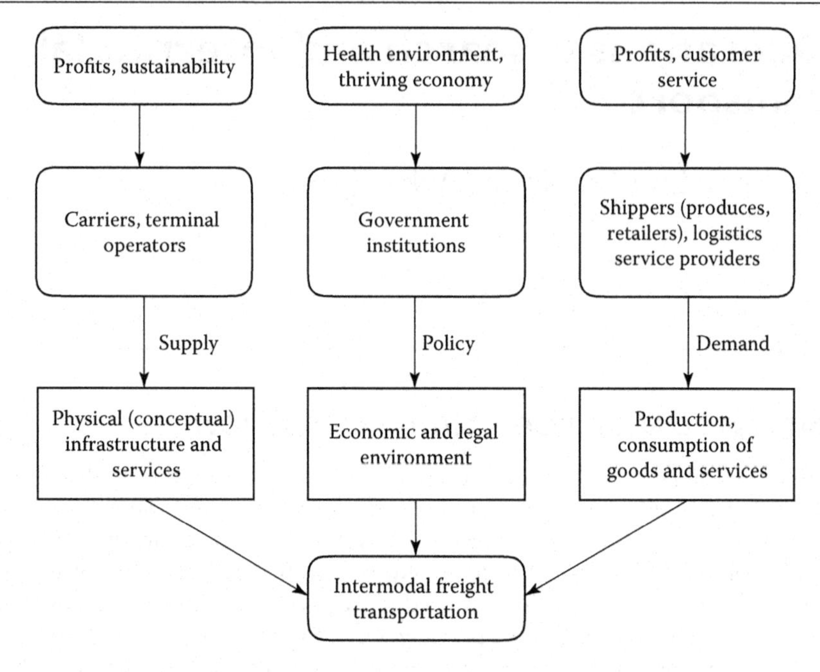

Figure 12.1 Stakeholders in intermodal freight transportation.

While *operations research* (OR) methods play a role in the decision making and planning of all three stakeholders mentioned earlier, the vast majority of research has focused on the supply side. Much of the planning of the supply of freight transportation services corresponds to the design of a network of terminals through which customer freight is routed from its origin to its destination. Different terminals can serve multiple purposes, including facilitating the transfer of freight between modes (e.g. intermodal terminals) and/or the consolidation of freight from multiple customers into a single vehicle. Along with determining the number of terminals and a location for each, a carrier must also determine how freight will be transported between them. This includes determining the frequency and schedule (and potentially size) of transportation, the mode used and whether the carriers own assets should be used.

The methods for designing such a network are often categorised by the level of planning they inform. While these levels of planning are commonly encountered in the management literature, we will next identify and discuss each as it typically appears in the context of the supply side of intermodal freight transportation.

- *Strategic planning*. This level often focuses on a high-level plan of transportation capabilities and capacities over a multi-year horizon (e.g. 3–10 years). One example is a carrier seeking to expand its use of intermodal freight transportation by choosing a port near which they will locate a terminal.

Another example is an organisation setting a target for the percentage of their fleet that will run on alternative fuels at a certain point in the future.

- *Tactical.* This level often focuses on a more detailed plan of transportation capabilities and capacities over the nearer term (e.g. the next year). Continuing with the example of locating a terminal near a port, planning at this level could dictate how many trucks travel each week to and from the terminal near the port and other terminals in their network. Regarding the alternative fuel example, given a fleet of both alternative and traditionally fuelled vehicles, a tactical planning question is the allocation of those vehicles to moving freight between specific terminals in the network, as such allocations can impact their use.
- *Operational.* This level often focuses on a precise plan of transportation capabilities and capacities in the immediate term (e.g. the next week). Again, and continuing the example of a terminal near a port, planning at this level could dictate which driver on which day at which time will drive a truck to and from the terminal near the port. With respect to alternative fuels, operational planning could involve designing the route of an electric vehicle on a daily basis in such a manner that the need to recharge the battery is recognised.

Having discussed the planning of intermodal freight transportation systems, we next turn to a discussion of what OR is.

PRIMER ON OPERATIONS RESEARCH

OR, according to the Institute for Operations Research and the Management Sciences (INFORMS), one of the leading international organisations serving both academics and practitioners of operations research, is 'a discipline that deals with the application of advanced analytical methods to help make better decisions'. The premise of OR is that instead of making decisions based on intuition or 'hunches', they can be made via a rational, systematic and data-driven process. These processes typically bring together methods and tools from multiple disciplines, including computer science, mathematics and statistics.

Fundamentally, OR is a science offering a suite of methods for modelling a system wherein there is a decision-making problem to be solved, and then using that model and appropriate solution methods to derive a set of good decisions. Examples of problems approached with OR range from a hospital determining its staffing schedule for nurses, a distribution company determining which warehouses to open and from which warehouse each customer should be served, a manufacturing company designing its production line and an airline determining (dynamically) how much to charge for each plane ticket. To effectively use OR, one must first understand the system that is to be studied and, then, build a mathematical representation, a *model*. Such models often involve four components:

1 The *data* that (partially) defines the system, and that the current study will not alter. Such data can include the capacity of a transportation container, the freight volumes that are anticipated over a period of time, the time it takes for a container to be loaded onto a barge at an intermodal terminal and the carbon emitted by a truck driving from one terminal to another.
2 The *decisions* that can be made in order to impact how the system performs. In transportation applications, these decisions usually have a yes/no ('do I open this facility?') or quantity ('how many containers should I send from facility A to facility B') nature.
3 The *dynamics* of the system or the *restrictions* that govern when a set of decisions can be executed in practice. These restrictions can be a result of physical limitations such as a container having a fixed capacity in terms of the weight it can hold. Or, they can be a result of policy, for example the maximum number of daily driving hours for truck drivers, or operational rules such as a shipment having to be transported.
4 The key performance metric or indicator for the system by which the quality of a set of decisions is measured. Often, quality is measured along a single dimension such as monetary cost, enabling judgement of when one set of decisions is better than another. However, in other cases, quality is measured along multiple dimensions, such as monetary cost and some proxy for environmental impact, in which case the decision maker must weigh trade-offs in these measures when considering one set of decisions over another.

OR models of systems are often categorised based on whether they are prescriptive or predictive. *Prescriptive models* are those that yield the set of decisions that are best with respect to a KPI. *Predictive models*, on the other hand, predict the values of KPIs of interest given a set of decisions. Predictive methods are often used in situations wherein there is uncertainty in the data of the system or the relationships between agents in the system are difficult to represent mathematically. We will next give examples of each type of model (Figure 12.2).

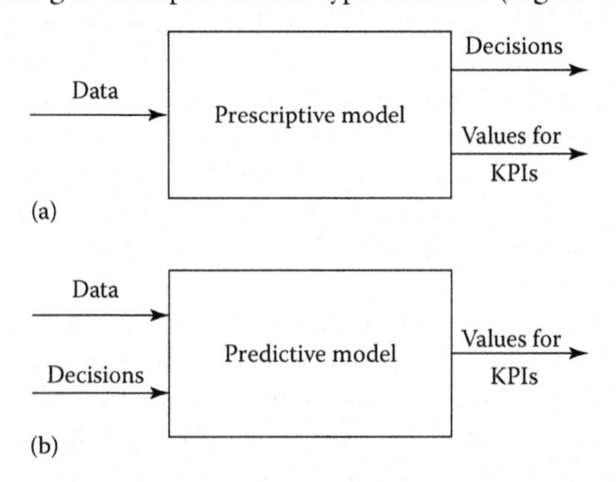

Figure 12.2 Inputs and outputs of two types of models.

PRESCRIPTIVE MODELS

One decision-making problem seen in intermodal transportation is that of a logistics service provider seeking to establish, through contracts with carriers, the weekly capacity for transportation between two locations that they can then offer to their customers. Carriers typically offer capacities in different units, with each unit having a different cost, and these units are often referred to as *bin types*. We will first present a model wherein it is presumed that the number of items and their sizes for each customer are known. Afterwards, we will present a model for settings wherein uncertainty in item sizes is recognised.

Returning to the components previously discussed, the primary decision for the logistics service provider is the number of bins of each type to commit to using. Fundamentally, the primary restriction is that the provider must be able to load each customer item into a bin, given that each bin has a fixed capacity. Thus, to accurately estimate whether a given number of bins of each type will provide the capacity needed to hold customer items, it is also necessary to decide the assignment of items to bins. With a single bin type, the quality of a set of decisions can be measured by the number of bins contracted. With multiple bin types, each with (potentially) different costs, the decisions are often measured in terms of how much they cost in total. This problem (and others like it) are often referred to as *bin packing* problems in the academic literature (Martello and Toth, 1990; Crainic et al., 2011; Baldi et al., 2012, 2014; Perboli et al., 2014).

Having formally defined the problem, the next step when using OR is to build a model of the problem. These models often take on a mathematical form and are referred to as mathematical programs. Such programs have four components, with each mapping directly to the aforementioned components.

We will discuss these components in the context of the previously described bin packing problem. To do so, we start with the *data* and let $T = \{1, 2, \ldots, t\}$ represent the set of different bin types, with each type $\tau \in T$ having capacity b_τ and cost f_τ. For example, bin type $\tau = 1$ may provide 20,000 lbs of capacity ($w_1 = 20{,}000$) at a cost of \$2,000 ($f_1 = \$2{,}000$). We then let $J^\tau = \{1, 2, \ldots, J\}$ represent the set of individual bins of type τ that can be acquired (via contract) and $J = \bigcup_{\tau \in T} J^\tau$ the set of all possible bins. We also let the set $I = \{1, 2, \ldots, I\}$ represent the different customer items, with each item $i \in I$ having size q_i. For example, customer item $i = 1$ may weigh 300 lbs ($q_1 = 300$).

We then define the *decision variables* that represent the decisions that can/must be made. Decisions that have a yes/no nature are usually modelled with binary variables (they can only take on the value zero or one). Decisions regarding quantities are usually modelled with integers (they typically can only take on values 0, 1, 2,...) or (non-negative) continuous variables. In the foregoing example, there are two sets of decisions: whether an individual bin of a given type is to be used and in which bin each customer item should be loaded. Each set of decisions is modelled with a different type of decision variable. The first type (labelled y_j^τ , $j \in J^\tau$) is required to take on the value zero or one and models whether bin j of type τ is to be used. Then, to model the assignment of items to bins, the second type (labelled x_{ij}, $i \in I, j \in J^{\tau}$) is required to take on the value zero or one, and models the yes or no decision of whether customer item i is to be loaded into bin j of type τ.

The third component represents the *constraints* that limit the values that the decision variables can take on due to problem restrictions. The decision variables then appear within mathematical equations or inequalities that ensure the values of the decision variables do not imply a set of decisions that violate the problem restrictions. For example, the restriction imposed by a bin's capacity can be expressed as $\sum_i \in I q_i x_{ij} \leq b_{\tau}$, $j \in J^{\tau}$. In this constraint, the expression $\sum i \in I\ q_i x_{ij}$ represents the total number of pounds loaded into bin j. However, it is also necessary to ensure that bins that are used are paid for, which can be modelled by adjusting the previous constraint to $\sum_{i \in I} q_i x_{ij} \leq b_{\tau} y_j^{\tau}$. Whereas the previous constraint models a physical restriction (bin capacity), the constraint $\sum_{j \in J} x_{ij} = 1$, $i \in I$ is needed to model the operational restriction that each item must be put into a bin.

Finally, the fourth component models the *objective function* that maps the values of decision variables to a single quantitative measure, assuming quality is measured along a single dimension. When quality is measured along multiple dimensions, multiple objective functions are used, one for each dimension. In our example, the objective function is $\sum_{j \in J} f_{\tau} y_j^{\tau}$ and measures the total expenditure on bins.

As such, a mathematical program for the capacity planning problem for the logistics service provider is as follows:

$$\text{minimize} \sum_{j \in J} f_{\tau} y_j^{\tau}$$

$$\sum_{i \in I} q_i x_{ij} \leq b_{\tau} y_j^{\tau}, j \in J^{\tau},$$

$$\sum_{j \in J} x_{ij} = 1, i \in I,$$

$$y_j^{\tau} \in \{0,1\}\ j \in J^{\tau},$$

$$x_{ij} \in \{0,1\}\ i \in I, j \in J^{\tau}.$$

Such a mathematical program, coupled with values for the parameters f_{τ}, b_{τ} and q_i (completing the data component) can then be solved with off-the-shelf software, which in turn yields values for each of the decision variables and the costs incurred by executing those decisions. Then, those values can be used to derive a set of decisions that can be implemented in practice.

This mathematical program is known as an *integer program* (IP) as the variables are required to take on integer (actually binary) values, and the objective function and constraints are linear in the decision variables. After solving a linear program (linear objective function and constraints and decision variables that can take on fractional values), mathematical programming solvers can produce sensitivity analysis reports as well as the values of decision variables. These reports have a 'what-if' nature, and answer such questions as, 'how does the optimal cost change when the value b_j changes?'. However, in some transportation applications, non-linear functions can appear either in the objective or a constraint. This is particularly prevalent in problems involving congestion, wherein increases in volumes can have an exponential impact.

When building such mathematical programs, it is common to model only a subset of the decisions that need to be made and restrictions that must be observed. In our bin packing example, there are also decisions to be made regarding how items are loaded into a bin. And the model discussed previously measures capacity solely with respect to weight carried, whereas the volume of a bin also limits what it can carry. Finally, there can be situations (such as with chemicals) where different items cannot be placed in the same bin. In these (and other) cases, the planner must decide whether to include loading decisions in the mathematical model or to simply adjust the solution prescribed by the model after the fact.

The previous mathematical program could be classified as *deterministic*, as it presumed that we knew the values for the parameters b_τ, f_τ and q_i. Namely, that we knew the capacity and cost of each bin as well as the weight of each customer item. In a situation where bins are acquired for weekly use via contract, knowing the capacity and cost of each bin is likely reasonable. However, presuming the values for q_i are known with certainty when the contract is negotiated, implies that each customer item will weigh the same amount each week and that one knows what that amount is. For medium- and long-term contracts, making such an assumption may lead to contracting for too many (or too few) bins.

In many intermodal freight transportation applications, parameter values are not known with certainty when a planning model is solved, either because of measurement error or simply because they cannot be presumed known. Many such decision-making problems are assisted by solving *two-stage stochastic programs*. Such mathematical programs are built on the premise that there are a fixed set of potential states of nature (e.g. 'low', 'medium' and 'high' demand) that can occur; these states of nature are often referred to as *scenarios* and it is presumed that a probability of each scenario occurring can be estimated. These mathematical programs also model a decision-making process wherein some decisions must be made before the value of stochastic parameters are known (*first-stage* decisions), and other decisions can only be made after those values are known (*second-stage* decisions). Returning to our bin packing example, instead of the logistics service provider determining the capacity it acquires via contract, based on a single estimate of the demands it expects from its customers, we can consider a model wherein uncertainty in the size of customer items is recognised (Crainic et al., 2016a). To that effect, we assume a set of scenarios, Ω, of potential customer demands and a probability p_ω of scenario $\omega \in \Omega$ occurring. Specifically, we now model customer demands with the parameter q_i^ω, $i \in I, \omega \in \Omega$. For example, now, in the 'low' demand scenario ($\omega = 1$), we model that item 1 will weigh 100 lbs ($q_1^1 = 100$), but in the 'high' demand scenario ($\omega = 3$), that same item will weigh 1000 lbs ($q_1^3 = 1000$). Of course, the model can be instantiated with as many scenarios as the planner feels is necessary to capture the potential states of nature that he or she will encounter.

While the choice of bins to acquire via contract is modelled as being made before demand is known, it is common in these models to recognise that extra capacity can be acquired (at a higher cost) on a 'spot' market when necessitated by customer demand. Such actions are often referred to as *recourse* or how an organisation reacts to the realisation of uncertain parameter values. While we will focus on the acquisition of extra capacity when needed, in some settings, recourse could include the leasing/resale of capacity acquired via contract that is in excess of customer demand.

Regarding the acquisition of extra capacity, we presume a set Θ of bin types that can be acquired on an on-demand basis, and index the set of 'spot' bins of each type by K^{θ}. Associated with bin k of type θ we associate a cost f_{θ} and capacity b_{θ}. For notational convenience, we let $K = \bigcup_{\theta\in\Theta} K^{\theta}$.

As such, in our mathematical program we can categorise our decisions into first- and second-stage decisions. The first-stage decisions are the y_j^{τ} variables described in the previous model. Then, there are two sets of second-stage decisions (those that can be made after demand is observed each week). The first represents the decisions regarding 'spot' bins acquired and are modelled with the variable $y_k^{\theta\omega}$ (a binary variable) that indicates whether bin k of type θ is acquired in scenario ω. The second represents the fact that the assignment of items to bins need not be made until item sizes are known. It is modelled with the binary variable x_{ij}^{ω}, $j \in J \cup K$, representing whether in scenario ω customer item i is loaded into bin j that is acquired either via contract ($j \in J$) or on the spot market ($j \in K$). Ultimately, the objective of many two-stage stochastic programs is to minimise the sum of costs associated with first-stage decisions and the expected cost associated with recourse decisions. Now, to determine how many bins of each type to acquire via contract, we can solve the following model:

$$\text{Minimize} \sum_{j\in\mathcal{J}} f_{\tau} y_j^{\tau} + \sum_{\omega\in\Omega} p_w \sum_{k\in\mathcal{K}} f_{\theta} y_k^{\theta\omega}$$

$$\sum_{i\in\mathcal{I}} q_i x_{ij}^{\omega} \le b_{\tau} y_j^{\tau}, j \in \mathcal{J}^{\tau}, \omega \in \Omega, \tag{12.1}$$

$$\sum_{i\in\mathcal{I}} q_i x_{ik}^{\omega} \le b_{\theta} y_k^{\theta}, k \in K^{\theta}, \omega \in \Omega, \tag{12.2}$$

$$\sum_{j \in J} x_{ij}^{\omega} + \sum_{k \in K} x_{ik}^{\omega} = 1, i \in \mathcal{I}, \omega \in \Omega, \quad (12.3)$$

$$y_j^{\tau} \in \{0,1\} \; j \in J^{\tau},$$

$$y_k^{\theta\omega} \in \{0,1\} \; k \in K^{\theta}, \omega \in \Omega$$

$$x_{ij}^{\omega} \in \{0,1\} \; i \in \mathcal{I}, j \in J \cup K, \omega \in \Omega.$$

Note now there are two constraints regarding bin capacity, one (Constraint 12.1) involving bins that are acquired via contract and the other (Constraint 12.2) involving bins acquired via the spot market. Also, note that now, in another constraint (12.3), an item may be assigned either to a bin acquired via contract or one acquired via the spot market.

Finally, we note that while instances of the two mathematical programs just presented can theoretically be solved directly by off-the-shelf software, many instances are too difficult for standard software to solve in a run-time that is acceptable for planning in practice. For such instances, a planner may draw upon two other types of solution approaches: (1) heuristics, which are algorithms that are designed to be very fast, but are not guaranteed to produce an optimal solution; and (2) solution methods, such as Benders decomposition, that are guaranteed to produce an optimal solution, and are designed either for the specific problem at hand (e.g. stochastic bin packing) or the problem class (two-stage stochastic programs). The advantage of each type of approach is that they are typically more effective than off-the-shelf software. The disadvantage is that they often involve a custom implementation.

Recall that we identified three stakeholders in intermodal freight transportation: shippers, carriers and government institutions. Each stakeholder will likely use a prescriptive model in the course of planning their operations. We will next give examples of each.

Shipper

While shippers demand transportation services, they often supply products to consumers. For a retailer, critical questions include how often and how much stock should be delivered from warehouses to store locations. These inventory policies have a direct effect on the retailers demand for transportation services. Inventory policies are often determined based on the speed and reliability of the transportation used, both of which can vary across modes. The classic textbook on inventory models remains Silver et al. (1998).

Carrier/terminal operator

The example discussed earlier illustrates how a carrier can use a prescriptive model to determine capacity acquisition. A second example is the ***hub location*** or ***hub network design*** class of problems (O'Kelly, 1986; Campbell et al., 2002;

Alumur and Kara, 2008; Farahani et al., 2013). Such models are used to assist the selection of one or several terminals, among those making up the system, as major (intermodal) hubs, and the selection of the transportation services to connect the other, regional, terminals to the hubs. The flows of services and vehicles will then converge upon those hubs, where classification, consolidation and transfer operations will be performed efficiently, also providing the opportunity for economies of scale in inter-hub transportation. Consolidation-based carriers, for example railroads, less-than-truckload motor carriers, intermodal container-transportation carriers and express couriers, operate according to such a hub-based structure.

A third, classical example of a prescriptive model used by a carrier is the *service network design* model (Crainic, 2000; Crainic and Kim, 2007; Crainic et al., 2016b; Hewitt et al., 2010; Zhu et al., 2014). This model assists carriers in determining the transportation services they will offer between terminals in their network, the terminal workloads and the routing of freight within the service network thus determined. The model can include such decisions as frequencies, resources assigned to support the execution of a transportation service and their management and the mode used. Instead of frequencies, scheduled service network design models will prescribe the dispatch times for each transportation service at each of the stops on its route.

Several surveys address carrier/terminal operator issues and models, including Bektaş and Crainic (2008), Crainic and Kim (2007) and Macharis and Bontekoning (2004) for intermodal; Cordeau et al. (1998) for rail; Christiansen et al. (2004, 2007) for maritime transportation; Powell et al. (2007) for fleet management; Bierwirth and Meisel (2010), Steenken et al. (2004) and Stahlbock and Voß (2008) for port terminal operations.

Government institution

In many areas, the operation of transportation infrastructure is managed by governmental employees (e.g. rail operators). Such an agency may use a prescriptive model to plan the strategic development of the network (Crainic et al., 1990b; Andersen and Christiansen, 2009) or to schedule the workload of each operator. Alternately, governments may use a prescriptive model to plan passenger-oriented transportation with an understanding that it relies on physical infrastructure that is also used for freight. For example, a regional transportation authority may develop a schedule for passenger trains on a rail line with the understanding that a certain fraction of that line's capacity must be reserved for freight trains.

PREDICTIVE MODELS

In the previous section, we presented two models that can be solved in order to assist a logistics provider that is trying to determine how much capacity to acquire. In particular, those models can be solved to yield the number of containers or bins to acquire via contract. In this section, we turn our attention to models that instead of prescribing decisions, take them as input, and predict the values of performance metrics of interest. As an example, we consider loading and unloading ships at an intermodal container terminal and the question of how many container cranes should be purchased to ensure ships do not, on average, wait more than a given amount of time.

One can answer such a question in many ways. One method would be through a simple capacity analysis wherein one assumes that when the utilisation of a system is less than 100%, then no ships will wait once the system reaches a steady state. However, if the utilisation of the system is greater than 100%, then the system will never reach a steady state and each ship will wait longer than the last. Utilisation of the system can then be calculated with the formula $u = p/ma$, where p represents the number of hours needed to load/unload a ship, a is the number of hours that pass between ships arriving to be loaded/unloaded and m is the number of container cranes. With this formula, one can simply choose a value of m such that $u < 100\%$.

However, such an analysis makes many simplifying assumptions regarding the dynamics of the system. In particular, it presumes that there is no variability. For example, it presumes that every ship requires exactly p hours to load/unload, and every ship arrives to be loaded/unloaded exactly a hours after the last ship arrived. In the queueing literature, p is often referred to as the processing time and a as the inter-arrival time. To predict the average amount of time a ship would have to wait without these assumptions, a planner can draw on results from the field of queueing theory (Gross and Harris, 2008). In particular, if a planner is willing to assume that both the number of hours required to load a ship and the number of hours that pass between ships arriving follow a given probability distribution, one can employ the following formula to predict the average waiting time:

$$T_q = \frac{p}{m} * \frac{u^{\sqrt{(2*(m+1))}-1}}{1-u} * \frac{CV_a^{\,2} + CV_p^{\,2}}{2} \tag{12.3}$$

In this formula, u, p, a and m are as discussed previously, whereas CVa and CVp represent the coefficients of variation of the probability distributions used to model variations in the processing and inter-arrival time (respectively). Note that while such a formula is only an approximate estimate, empirical studies have shown that it is quite accurate, particularly when the probability distributions are the exponential distributions.

That said, Equation 12.3 is based upon many assumptions as well. For example, it presumes that the processing time for each ship follows the same probability distribution. It also assumes that every container crane is always available, and that they all have the same capabilities. The equation is also based on the presumption that each ship will wait as long as it must for a crane to become available, and that there is no limit on how many ships wait to be loaded/unloaded. Finally, the equation is based on the presumption that only the crane is necessary for loading and unloading a ship, when in fact there may be other resources that must participate in the process that have their own processing times and availabilities.

While the literature on queueing theory contains models for many situations, at some point, a planner must resort to computer simulation models (Law et al., 1991). Much like how automotive engineers can build simulation models to understand the impacts of their design decisions on the aerodynamics of a vehicle, a logistics planner can build a simulation model of an intermodal terminal to understand how the number of cranes impacts average waiting time. One advantage of these models is their flexibility; modern simulation software can make it easy for a planner to model very complicated system dynamics. That said, simulation models require development and the necessary software to do so.

Finally, a disadvantage associated with predictive models in general is that finding the best set of decisions is a trial-and-error process. For our crane example, the 'optimal' number could be found by enumerating over the potential numbers of cranes that could be acquired, running a simulation model for each number and choosing the smallest number that yields an average waiting time that is less than the threshold. However, for more complex, and potentially multilayered sets of decisions, there may be too many combinations to enumerate. For example, in an intermodal terminal, there may be more berths than cranes, and thus the planner must determine both the number of cranes and the schedule for when each crane is at each berth. Therefore, this schedule has a great impact on the waiting time for a given number of cranes, and thus, the two decisions are interrelated.

That said, predictive models are often used by each of the stakeholders mentioned earlier (shippers, carriers and government institutions), as illustrated in the survey by Crainic et al. (2016c). We will next discuss an example of each.

Shipper

Recalling our example of a retailer using a prescriptive model, what often drives their determination of an inventory policy is a predictive (e.g. forecasting) model of the anticipated demand in each store for their products. A good primer on forecasting is Hyndman and Athanasopoulos (2014).

Carrier/terminal operator

The example discussed earlier illustrates how a terminal operator could use a predictive model to configure a port. Another common use for simulation-based predictive models is to evaluate the robustness of a plan that was determined under very strong assumptions regarding system dynamics and behaviour. For example, a carrier operating in a same-day delivery or just-in-time manufacturing environment will have little time to route freight from its origin to its destination through a network of terminals. As such, the routing of freight through that network requires a high degree of synchronisation between different services. The carrier may use a prescriptive service, a network design model to determine services (including their dispatch times) under the assumption that travel times are known and constant and then a simulation model to evaluate the use of those services when stochasticity in travel times is experienced.

Government institution

Governments at various levels (e.g. European Union, national, regional and municipal), as well as many financing institutions (e.g. the World Bank), often dictate policies and make decisions on investments for new infrastructure or the upgrade of existing infrastructure that will impact the behaviour and performance of the transportation system. Hence the need to evaluate these probable impacts before the policies are set or the investments are approved. Prescriptive models used for such evaluations simulate the behaviour of the system through an optimisation procedure that assigns estimations of future demands to a representation of the future state of the transportation network (Crainic et al., 1990a; Crainic and Florian, 2008).

Governments also often dictate policies which impact how physical infrastructure can be used. For example, a regional transportation authority may dictate the degree by which a rail line is shared among trains that are carrying freight and those that are carrying passengers and/or which type of train has priority. Such a change in policy can impact the decision making and satisfaction of both (multiple) carriers and passengers. Thus, to evaluate such a change, a governmental agency may use a predictive model that simulates the behaviour of these different players.

PERSPECTIVES AND CONCLUSIONS

This chapter focused on OR and intermodal freight transportation. Intermodal freight transportation involves complex systems with various stakeholders, decision levels and objectives. OR provides the methods – modelling, algorithmic and analytic frameworks – to represent and analyse such systems and to assist decision making.

The chapter started with a brief description of intermodal transportation, which yielded the identification of the main stakeholder classes, and an OR primer introducing the main components of OR models. We then focused, first, on prescriptive models and then on descriptive ones. In both cases, we first detailed the general structure of such models, using intermodal transportation cases, and then illustrated a few important problems and models for each of the stakeholder classes.

The chapter aimed to be informative and to provide a starting point for the understanding and application of OR to planning and managing intermodal transportation systems. This field is extremely rich in both applications and fascinating research challenges. We hope the interested reader will follow some of the leads offered by this chapter and continue exploring the field.

REFERENCES

Alumur, S., Kara, U. Network hub location problems: The state of the art. *European Journal of Operational Research*, 190 (1): 1–21, 2008.

Andersen, J., Christiansen, M. Designing new European rail freight services. *Journal of the Operational Research Society*, 60 (March): 348–360, 2009.

Baldi, M. M., Crainic, T. G., Perboli, G., Tadei, R. Branch-and-price and beam search algorithms for the variable cost and size bin packing problem with optional items. *Annals of Operations Research*, 222: 125–141, 2014.

Baldi, M. M., Crainic, T. G., Perboli, G., Tadei, R. The generalized bin packing problem. *Transportation Research Part E: Logistics and Transportation Review*, 48 (2): 1205–1220, 2012.

Bektaş, T., Crainic, T. G. A brief overview of intermodal transportation. In Taylor, G. D., editor, Logistics Engineering Handbook, Chapter 28, pp. 1–16. CRC Press, Boca Raton, FL, 2008.

Bierwirth, C., Meisel, F. A survey of berth allocation and quay crane scheduling problems in container terminals. *European Journal of Operational Research*, 202 (3): 615–627, 2010.

Campbell, J. F., Ernst, A. T., Krishnamoorthy, M. Hub location problems. In Drezner, Z., Hamacher, H., editors, *Facility Location: Application and Theory*, pp. 373–407. Springer Verlag, Berlin, 2002.

Christiansen, M., Fagerholt, K., Nygreen, B., Ronen, D. Maritime transportation. In Barnhart, C., Laporte, G., editors, *Transportation*, volume 14 of *Handbooks in Operations Research and Management Science*, pp. 189–284. North-Holland, Amsterdam, 2007.

Christiansen, M., Fagerholt, K., Ronen, D. Ship routing and scheduling: Status and perspectives. *Transportation Science*, 38 (1): 1–18, 2004.

Cordeau, J. F., Toth, P., Vigo, D. A survey of optimization models for train routing and scheduling. *Transportation Science*, 32 (4): 380–404, 1998.

Crainic, T. G. Network design in freight transportation. *European Journal of Operational Research*, 122 (2): 272–288, 2000.

Crainic, T. G., Florian, M. National planning models and instruments. *INFOR*, 46 (4): 81–90, 2008.

Crainic, T. G., Florian, M., Guélat, J., Spiess, H. Strategic planning of freight transportation: STAN, an interactive-graphic system. *Transportation Research Record*, 1283: 97–124, 1990a.

Crainic, T. G., Florian, M., Léal, J. E. A model for the strategic planning of national freight transportation by rail. *Transportation Science*, 24 (1): 1–24, 1990b.

Crainic, T. G., Gobbato, L., Perboli, G., Rei, W. Logistics capacity planning: A stochastic bin packing formulation and a progressive hedging meta-heuristic. *European Journal of Operational Research*, 253 (2): 404–417, 2016a.

Crainic, T. G., Hewitt, M., Toulouse, M., Vu, D. M. Service network design with resource constraints. *Transportation Science*, 50 (4): 1380–1393, 2016b.

Crainic, T. G., Kim, K. H. Intermodal transportation. In Barnhart, C., Laporte, G., editors, *Transportation*, volume 14 of *Handbooks in Operations Research and Management Science*, Chapter 8, pp. 467–537. North-Holland, Amsterdam, 2007.

Crainic, T. G., Perboli, G., Rei, W., Tadei, R. Efficient lower bounds and heuristics for the variable cost and size bin packing problem. *Computers and Operations Research*, 38 (11): 1474–1482, 2011.

Crainic, T. G., Perboli, G., Rosano, M. Simulation of freight intermodal transportation systems: A taxonomy. Cirrelt publication, Centre interuniversitaire de recherche sur les réseaux d'entreprise, la logistique et le transport, Université de Montréal, Montréal, QC, Canada, 2016c.

Farahani, M. E., Hekmatfar, M., Arabani, A. B., Nikbakhsh, E. Hub location problems: A review of models, classification, solution techniques, and applications. *Computers and Industrial Engineering*, 64 (4): 1096–1109, 2013.

Gross, D., Harris, C. M. *Fundamentals of Queueing Theory*. Wiley, New York, 2008.

Hewitt, M., Nemhauser, G., Savelsberg, M. Combining exact and heuristic approaches for the capacitated fixed-charge network flow problem. *INFORMS Journal on Computing*, 22 (2): 314–325, 2010.

Hyndman, R. J., Athanasopoulos, G. *Forecasting: Principles and Practice*. OTexts, 2014.

Law, A. M., Kelton, W. D. *Simulation Modeling and Analysis*, volume 2. McGraw-Hill, New York, 1991.

Macharis, C., Bontekoning, Y. M. Opportunities for OR in intermodal freight transport research: A review. *European Journal of Operational Research*, 153 (2): 400–416, 2004.

Martello, S., Toth, P. *Knapsack Problems: Algorithms and Computer Implementations*. Wiley, New York, 1990.

O'Kelly, M. E. The location of interacting hub facilities. *Transportation Science*, 20: 92–106, 1986.

Perboli, G., Gobbato, L., Perfetti, F. Packing problems in transportation and supply chain: New problems and trends. *Procedia: Social and Behavioral Sciences*, 111: 672–681, 2014.

Powell, W. B., Bouzaïene-Ayari, B., Simaõ, H. P. Dynamic models for freight transportation. In Barnhart, C., Laporte, G., editors, *Transportation*, volume 14 of *Handbooks in Operations Research and Management Science*, pp. 285–365. North-Holland, Amsterdam, 2007.

Silver, E. A., Pyke, D. F., Peterson, R. et al. *Inventory Management and Production Planning and Scheduling*, volume 3. Wiley, New York, 1998.

Stahlbock, R., Voß. S. Operations research at container terminals: A literature update. *OR Spectrum*, 30 (1): 1–52, 2008.

Steenken, D., Voß, S., Stahlbock, R. Container terminal operation and operations research: A classification and literature review. *OR Spectrum*, 26 (1): 3–49, 2004.

Zhu, E., Crainic, T. G., Gendreau, M. Scheduled service network design for freight rail transportations. *Operations Research*, 62 (2): 383–400, 2014.

Chapter 13

Environmental aspects of intermodal transport

Erik Fridell

INTRODUCTION

It is good practice for an environmental assessment of freight transportation systems to allocate the impact to the freight being transported. This means that the consequences of the whole system, for example emissions, are calculated together with the transport work, and the two are then related. Transport work should then be calculated as the mass of goods multiplied by the transport distance (unit tkm), which allows for a comparison of different transport modes. Another common practice is to look at the impact per transported twenty-foot equivalent units (TEUs) and distance (TEU-km), sometimes divided in empty and loaded containers. This makes comparison less certain, since the amount of cargo in the containers may vary significantly. The allocation of the impact to the cargo could mean that empty positioning of cargo carriers, a very frequent activity for containers, is not assigned a specific environmental impact, but rather the effect of this traffic is allocated to the pre-movement or subsequent movement of goods. Further, for vehicles not moving but still having an impact on the environment, for example a ship at anchor, the effects should be associated with the movement of goods. According to this point of view, construction, maintenance and operation of infrastructure; production and scrapping of vehicles and vessels; upstream effects from fuel production also should be considered and allocated to the transport work. The opposite view would lead to there being environmental problems, for example emissions, not assigned to the actual transport work but to other entities (headquarters, authorities, etc.), thus making it difficult to optimise transport systems from an environmental point of view.

Intermodal transport from an environmental point of view is not an extensively studied subject. Winebrake et al. (2008) present a tool for calculating transport time, cost and carbon dioxide (CO_2) emissions for different intermodal transport alternatives and show with examples that there is a large variation in emissions. Kreutzberger et al. (2003) conclude in a review that often there are environmental advantages with intermodal transport that uses rail or sea modes, especially concerning CO_2 emissions. There is often a general assumption that intermodal transport is more environmentally friendly than unimodal (the latter meaning road) (Lowe, 2005).

This chapter looks at the environmental aspects of intermodal transport. The focus is on container transportation, but the analysis would be similar for other types of cargo carriers. This chapter concerns the actual movement of goods. It does not consider the environmental issues related to infrastructure and the production of the containers or the vehicles themselves. However, the movement of goods (where typically the combustion of fossil fuels is the culprit) poses the largest impact on the environment and health risks from a freight transport system. Furthermore, the chapter does not address an analysis of air transport of intermodal goods.

IMPACT ON THE ENVIRONMENT

A lot can be said about the different environmental concerns related to intermodal transport. Today, the contribution to extensive global warming through the emissions of greenhouse gases, most importantly carbon dioxide (CO_2) but also methane, nitrous oxide and other substances, is most in focus. Emission of CO_2 is the consequence of the combustion of fuel containing carbon such as gasoline, diesel oil, natural gas and biofuels. CO_2 from biofuels (biogenic) is usually not considered a contributor to global warming, because CO_2 was taken from the air to grow the biomass. Significant methane emissions can come as a 'slip' while combusting natural gas. It is common to present the emission of greenhouse gases as CO_2 equivalents where the lifetime in the atmosphere and optical properties are considered. When doing this, the time perspective used must be presented (normally 20 or 100 years). However, for transportation using fossil fuels, CO_2 usually dominates. Further, fuel combustion will also lead to increased levels of particulate matter (PM), which have a complex climate impact.

The total emission of CO_2 equivalents in the European Union (EU-28) countries in 2014 was about 4 billion t (excluding international aviation and international shipping) (Eurostat, 2016a). About 22% of these emissions come from fuel combustion in the transport sector of which road traffic contributes 95%, domestic aviation and domestic navigation about 2% each and railways about 1%.* In addition, international navigation and international aviation in EU-28† contributes 0.27 and 0.14 billion t of CO_2 equivalents, respectively.

The emission of air pollutants is another concern. There are a number of substances that contribute to environmental problems and to health risks. Further,

* This number does not include emissions of greenhouse gases from electricity generation for electric trains.

† In the statistics, this is measured as greenhouse gas emissions resulting from the combustion of the fuel sold within EU-28 for international aviation and shipping.

Table 13.1 Emissions of air pollutants in EU-28 in 2013, excluding international shipping and aviation, in kilotonnes

	Sulphur oxides	*Nitrogen oxides*	*Non-methane volatile organic compounds*	*Particulates <10 μm*
Total emissions	3400	8200	7000	1900
Road transport	5.8	3200	820	220
Non-road transport	79	550	130	30

secondary pollutants are formed in the atmosphere from chemical reactions often triggered by sunlight. Some of the most important air pollutants will be discussed here. Sulphur oxides are formed when sulphur in the fuel is oxidised in an engine. The amount emitted is thus proportional to the fuel combusted and the sulphur content therein. Sulphur oxides contribute to problems with the acidification of lakes and soil but may also lead to the formation of secondary particles, which give rise to health risks. Nitrogen oxides are formed as NO in the cylinders of the combustion engines from oxygen and nitrogen in the air at high temperatures and pressure. NO is gradually oxidised into NO_2. NO_x (the sum of NO and NO_2) poses health risks and contributes to acidification and eutrophication, as well as to the formation of secondary particles and ozone. Carbon monoxide (CO) is a poisonous gas formed from incomplete combustion. Hydrocarbons (HC) comprise a range of substances that pose various degrees of health risks (some are carcinogenic) and that contribute to the formation of secondary particles and ozone. PM is presently the emission most in focus regarding health concerns. PM can be emitted directly from an engine (primary particles), mostly in the form of carbon particles (soot). Further, a number of processes lead to particle growth and the formation of new particles (secondary particles) during the cooling and dilution of the exhaust in the air. The result is a complex mixture with different chemical compositions and different sizes. Table 13.1 gives an overview of the emissions of important air pollutants in EU-28 in 2013 (Eurostat, 2016b).

While the emissions to air of climate gases and air pollutants are the most urgent environmental issues associated with our transport system, other effects should also be considered. Shipping will of course have an impact on the marine environment through waste, oil spill, the emission of grey and black water, the transportation of invasive species and so on. Noise is a large problem with air transport, as well as with rail and road transport and in some cases, ports. Other aspects include barrier effects and land use.

The aspects discussed here are common for the transport system and cannot be said to be unique to intermodal transport. It is rather the magnitude of the effects that may be different, as will be exemplified in the next section.

SPECIFICS FOR DIFFERENT MODES OF TRANSPORT

Road transport

The first EU emission standards for heavy-duty road vehicles (Euro I) came in 1992, and the latest (Euro VI) came into force in 2013. The legislation applies to HC, CO, NO_x and PM and has been significantly tightened during the years (see Table 13.2). The standards are designed for measurement in an engine bench and are expressed in grams of emission per engine work (in kWh). Regarding CO_2 emissions and fuel consumption, there are no specific regulations. Although engines have gained in efficiency, the change over time has been moderate.

Fuel consumption and emissions will depend on the speed of the vehicle and the topography of the road. Further, fuel consumption per kilometre may be significantly higher in urban areas and in congested traffic, due to frequent stops and accelerations. Also, the mass of the cargo will have a significant influence on the fuel consumption. Table 13.3 shows a few examples.

For trucks, there are no fundamental differences between different cargo carriers. The load factor is the main parameter determining the emissions related to

Table 13.2 EU emission standards for trucks

Stage	*Date*	*CO (g/kWh)*	*HC (g/ kWh)*	*NOX (g/kWh)*	*PM (g/kWh)*
Euro I	1992	4.5	1.1	8	0.36
Euro II	1996	4	1.1	7	0.25
	1998	4	1.1	7	0.15
Euro III	2000	2.1	0.66	5	0.1
Euro IV	2005	1.5	0.46	3.5	0.02
Euro V	2008	1.5	0.46	2	0.02
Euro VI	2013	1.5	0.13	0.4	0.01

Source: Based on data from dieselnet.com.

Table 13.3 Fuel consumption for a truck with trailer per kilometre for different road types, gradients and cargo load

Road type	*Road gradient*	*Cargo load factor: Weight*	*Fuel consumption L/10 km*
Average road	±2%	50%	3.17
Rural road	±2%	50%	3.01
Urban road	±2%	50%	3.86
Average road	0	50%	2.71
Average road	±6%	50%	5.82
Average road	±2%	0.0%	2.27
Average road	±2%	100%	4.05

Source: Based on data from NTMcalc.org.

transport work and can vary significantly for intermodal transport systems as well as for unimodal systems. A high degree of positioning of empty containers will, of course, in general be detrimental to the environmental performance.

Diesel oil is the dominant fuel for road transport. In the EU, road diesel is fossil oil blended with a small percentage of biofuel and with low content of sulphur (max 10 ppm). There are also trucks that use liquefied petroleum gas and engines that have been developed for a range of other fuels, such as alcohols, compressed natural gas and dimethyl ether (DME). There are also ongoing development projects with electric trucks that obtain power from electrified roads.

Rail transport

The amount of cargo within the carriers should be considered in the calculations for intermodal transport with railways. Since 2006, there are emission regulations for diesel train engines within the EU that regulate emissions of CO, HC, NO_x and PM (see Table 13.4). However, the allowed emissions, measured as emissions per engine work (unit g/kWh) are significantly higher than for modern road vehicles. Train engines in Europe typically use diesel fuel with the same quality as that used for road vehicles. Electrical engines will show no emissions from the actual train; however, it is common practice to consider the emissions from the electricity production. This leads to very large differences depending on the source of electricity: hydro power and wind power will give very low emissions, while coal power will give large emissions (for CO_2 even larger than those from diesel engines). Nuclear and natural gas power give intermediate emissions. Table 13.1 shows an example of data for different electricity sources. It is common practice to use the electricity mix that is purchased by the transport provider. If this is 'green electricity' (from hydro and wind power), as for example is the case for the Swedish rail system, then the emissions will be very low. If no special type

Table 13.4 EU emission regulations for locomotive diesel engines

Category3	*Stage*	*Net power (kW)*	*Date*	*CO (g/kWh)*	*HC (g/kWh)*	*HC + NOX (g/kWh)*	*NOX (g/kWh)*	*PM (g/kWh)*
Rail car	IIIA	130 <P	2006	3.5		4.0		0.2
Locomotive	IIIA	130 < P <560	2007	3.5		4.0		0.2
Locomotive	IIIA	P > 560	2009	3.5	0.5[a]		6.0[a]	0.2
Rail car	IIIB	130 < P	2012	3.5	0.19		2.0	0.025
Locomotive	IIIB	130 < P	2012	3.5		4.0		0.025
Rail car	V	0 < P	2021[b]	3.5	0.19		2.0	0.015
Locomotive	V	0 < P	2021[b]	3.5		4.0		0.025

[a] HC = 0.4 g/kWh and NO_X = 7.4 g/kWh for engines of P > 2000 kW and D > 5 L/cylinder.

[b] Suggested date.

Table 13.5 Life cycle emissions of CO_{2e} for different electricity generation technologies

Electricity generation technology	*CO_{2e} life cycle emissions (g/kWh)*
Hydropower	4
Wind energy	12
Nuclear energy	16
Natural gas	469
Oil	840
Coal	1001

Source: 50th percentile values from Moomaw W. et al., *IPCC Special Report on Renewable Energy Sources and Climate Change Mitigation*, Cambridge University Press, Cambridge, 2011.

of electricity is purchased, it is normal to use the mix for the actual country. The emissions will therefore vary significantly between different countries.

Waterborne transport

Container ships and roll-on/roll-off ships (RoRo) are the types of ships most relevant for intermodal transport. There are also ships that combine container and RoRo cargo as well as ships that combine RoRo with passengers (RoPax ships). While container ships come in different sizes, from about 1,000 up to 20,000 TEU, and are used for trans-ocean shipping as well as for short sea shipping, RoRo ships are mostly used for short sea shipping. However, RoRo ships still come in a range of sizes and with different characteristics that are designed in light of the routes in which they are used. Ships for intermodal transport are typically used as freight liners and operate on schedules. Further, the demands on service often mean that these ships operate at high speed in relation to other ship types, such as tankers and bulk cargo ships.

Emission regulation for international shipping is decided within the International Maritime Organization (IMO). There are regulations covering emissions to air and water; however, the air regulations are less stringent compared with those for engines on land. The sulphur content in the fuel is regulated; worldwide the limit is now 3.5% and will be lowered to 0.5% by 2020. In special sulphur emission control areas (SECAs), the limit is 0.1%. Currently, these areas comprise the Baltic Sea, the North Sea, the English Channel and coastal areas outside the United States and Canada. There are also regulations for NO_X divided into tiers with Tier I applied to engines from 2000, Tier II from 2011 and Tier III from 2016. However, Tier III is only in effect for ships operating in special NO_X emission control areas (NECAs), at present only the coastal areas outside the United States and Canada. Further, the allowed NO_X emissions depend on the engine speed, allowing for higher emissions from slow speed engines. There are no specific regulations for PM, HC or CO for international shipping, although one of the objectives of sulphur regulations is to decrease the PM emissions.

Heavy fuel oil (HFO) with high viscosity, high content of polyaromatic hydrocarbons and high sulphur content (2%–3%) has been the dominating maritime fuel for the last several decades. This fuel is a relatively cheap residual product from refineries. Before it is injected into engines, this fuel must be heated. Marine gas oil (MGO), a fuel similar to road diesel but with a higher sulphur content (0.1%), is also used. There is also a range of different fuel mixes with viscosity and sulphur content in between these fuels. In the SECAs areas, the use of MGO increased after 2015. The regulations have also spurred an interest in other fuels, most notably liquefied natural gas (LNG). The use of LNG gives very low emissions of particles and sulphur and significantly lower NO_x emissions than for fuel oils. The choice of fuel has a significant influence on the emissions of sulphur, PM and trace elements such as various metals; the use of HFO is by far the worst choice in this respect.

A number of factors determine the environmental performance of a ship used in intermodal transport. Most notably, these include ship size, operating speed, load capacity utilisation, fuel quality, engine characteristics and whether or not after-treatment technologies are in place. Container ships for trans-ocean operation have become increased in size. One reason is the improved fuel efficiency, measured as fuel consumed per TEU-km of transport work, that comes with increased size.

The resistance in water for a ship, and thus the work needed from the engine, per unit of time, is approximately a cubic function of the speed. Thus, even small changes in speed can lead to large fuel savings or losses; for example, a shift from 16 to 14 knots would decrease the fuel consumption (per km) by over 30%. After the crisis in 2009, a large number of ships began operating at lower speeds (slow steaming) to save on the cost of fuel and in response to an overcapacity in the market.

From a ship owner's perspective, the load capacity utilisation of a container ship is usually seen as the fraction of the maximum possible TEUs that is being utilised. However, in line with allocating the environmental performance to the freight being transported, one should also consider the fill rate of the actual containers. This is an even more crucial factor for RoRo ships (Hjelle and Fridell, 2012). For a RoRo ship, it is common to look at the available lane metres and the fraction of these being used. However, if the deck is occupied by trucks (or other cargo carriers), one should consider the fraction of the deck that is used for cargo containers as well as the capacity utilisation of the cargo containers in order to allocate the emissions to the goods.

Inland waterway shipping is sometimes treated as a separate mode. The ships used on inland waterways such as canals and rivers are typically small and barges are common. These ships are usually operating at low speeds and are, in general, very fuel efficient. In Europe, there are special regulations on the emissions to air and the quality of the fuel that can be used in these ships.

SUMMARY OF EMISSIONS PER TONNE-KM

As can be understood from the earlier discussion, the emissions per tonne-km of transport work can vary significantly within, as well as between, different transport modes. Several tools can be used to calculate the environmental performance of transport chains; two examples are NTMcalc, through which emissions and fuel consumption can be calculated for all transport modes and also for passenger traffic, and ECOTransIT, which is based on a routing system. Figure 13.1 gives emissions of CO_2 per tonne-kilometre for a number of vehicles and vessels. These are calculated with data from NTMcalc and ECOtransIT. Note that the fill rate of the containers is the same for all cases. The values given in Figure 13.1 will depend on a number of factors with different uncertainties. The load factor is of course one of them. Since the same factor for the containers is used for all cases, this will not influence the comparison. However, for ships and trains, there is also the question of the utilisation of available space. For road traffic, the amount of empty legs will influence the overall results. In all cases, the positioning of empty containers will be an important parameter.

When assessing the climate impact of vehicles and vessels, the size and type of ship is important for the results, with container ships and larger ships showing better results than RoRo and smaller ships. For trains, the main factor is the energy form used. With 'green' electricity, the emissions become very small. For road traffic, there is a size dependence, but the emission standard (age of the truck) will only have a small influence. One could argue that it might be more difficult to obtain high load rates for larger vessels and vehicles compared with smaller ones, which would decrease the difference in the results for different sizes. The use of renewable fuels will radically change the results for emissions of greenhouse gases since biogenic CO_2 would not be counted. However, the production of biofuels may lead to significant emissions if it involves the heavy use of machinery and fertilisers as well as the transport of the fuels and/or raw

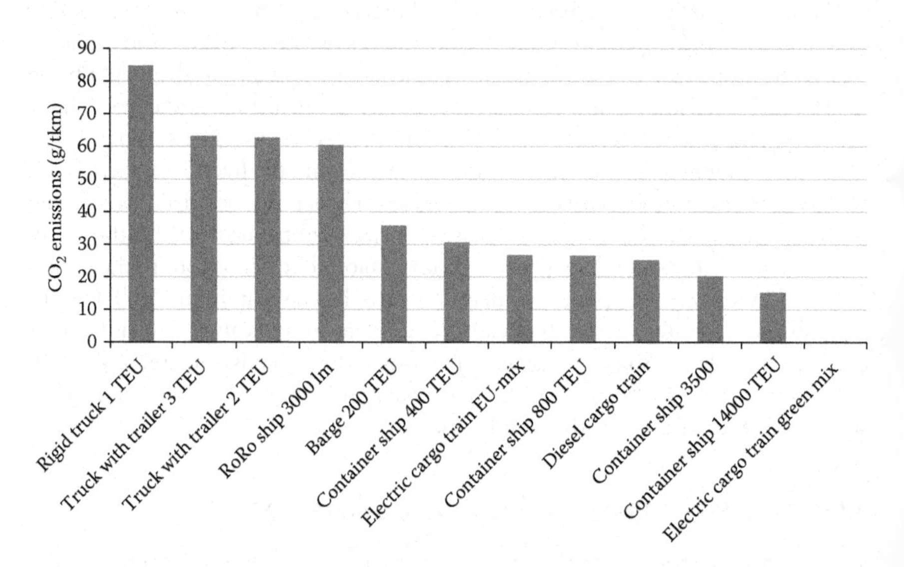

Figure 13.1 Typical emissions of CO_2 per tonne-kilometre for different vehicles/vessels assuming 7 t of cargo per TEU and load factors of 70%.

materials. For an overview of these upstream effects, see JEC (2014). Further, other greenhouse gases must be considered, such as methane and nitrous oxide. For other important emissions, such as PM and nitrogen oxides, the results will depend heavily on the emissions standard of the vehicle or vessel.

NODES

For an intermodal transport system, the nodes are, of course, essential. There are not many studies on the environmental impact associated with nodes; the reason being that it is expected to be small. The impact from the node should be added to the impact from the moving vessel or vehicle. As an example, let us consider a port. Environmental problems may include emissions to air from machinery and ships at berth; noise from engines, fans and cargo handling; climate impact from heating of buildings; pollution of water with oil and waste. These problems may be of importance for the port, especially if it is, as is common, located close to a city centre.

Stripple et al. (2016) analysed the Port of Gothenburg using the life cycle assessment (LCA) methodology when assessing construction maintenance and operation of the port. The emissions associated with the port were allocated to the amount of goods over quay (in tonnes moved). As an example, the CO_2 emissions associated with the port allocated in this way gave 0.3–5.0 kg CO_2 per tonne of goods over quay. The span reflects that the port contains different terminals handling different types of freight. For containerised goods, the range was 0.6–5.0 kg CO_2 per tonne of goods. Comparing this with the values for transportation given in Figure 13.1 shows that the emissions of greenhouse gases associated with the port correspond to a transport between 30 and 45 km for a truck or 70– 150 km for a container ship.

EXAMPLES

It is illustrative to compare an intermodal transport with a unimodal transport regarding environmental impact. In this chapter, we focus on emission to air of pollutants and greenhouse gases. It should be clear from the previous discussion that there are a number of choices regarding fuel, vessel size, emission standard, electricity type and so on that will have a significant impact on the results. All these choices cannot be covered here.

As an example, we will look at transport from Jönköping in Sweden to Mainz in Germany, a distance of about 1300 km by road including a passage across the Baltic Sea. This example is chosen because it illustrates different competing modes of transport. Consider the following four basic examples:

1. A truck picks up goods in Jönköping and drives to Mainz to deliver the goods.
2. A truck picks up goods in Jönköping, drives to the train terminal in Gothenburg to deliver the goods and a train takes the goods to Mainz.

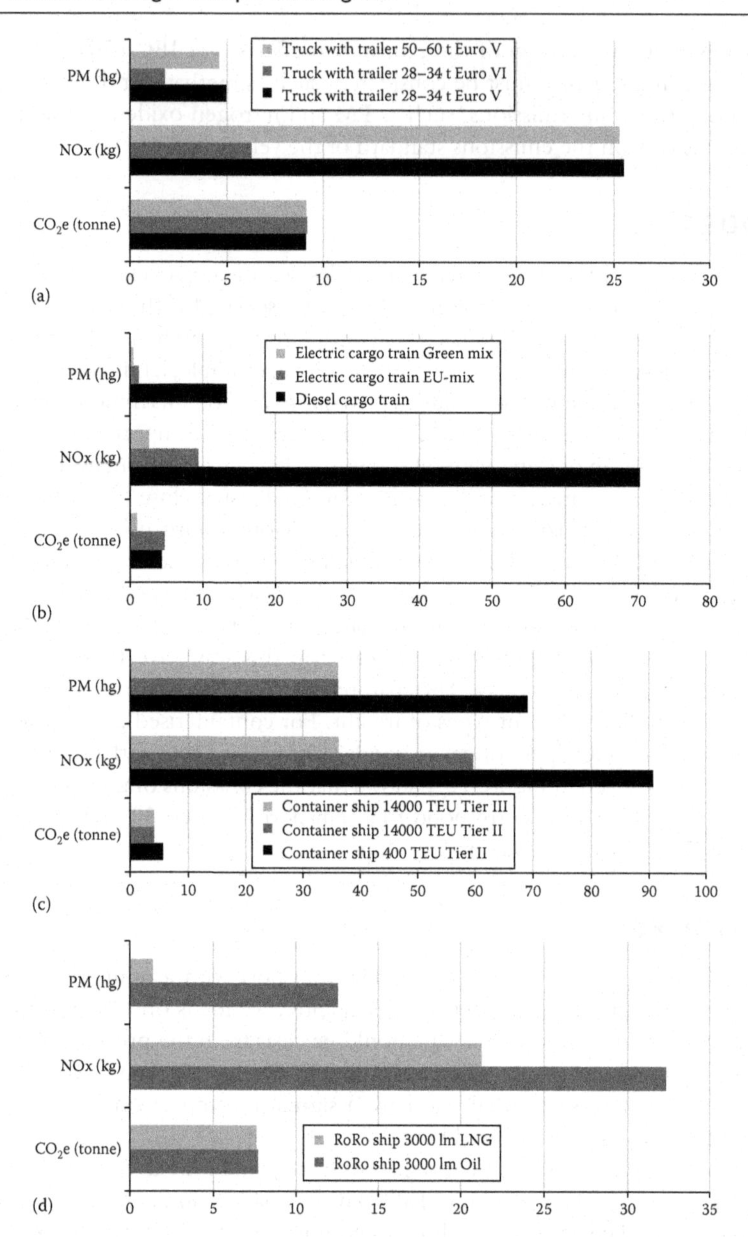

Figure 13.2. Emissions of CO_2, NO_x and PM for different unimodal and multimodal transport chains for the transport of 100 t of goods between Jönköping in Sweden and Mainz in Germany. (a) Unimodal truck chain, (b) truck-train chain, (c) truck-container–ship-barge chain, (d) truck and RoRo ship chain. See main text for descriptions of the chains. Data for emissions are taken from NTMcalc and ECOTransIT. A load utilisation factor of 70% is assumed for all cases.

3. A truck picks up goods in Jönköping and drives to the Port of Gothenburg where the goods are loaded onto a container ship; the ship takes the goods to Rotterdam where they are reloaded onto a barge for Mainz.
4. A truck picks up goods in Jönköping and drives to Trelleborg, enters a RoRo vessel for Sassnitz and then drives to the recipient in Mainz.

Within each of these basic examples, a few parameters have been varied in order to illustrate their impacts:

1. Trucks with different emissions standards and different sizes
2. Diesel trains and electric trains with different sources of electricity
3. Different sizes of container ships
4. RoRo ship with LNG and fuel oil

Assessments have been made for greenhouse gases (as CO_2 equivalents), PM and nitrogen oxides. The results are illustrated in Figure 13.2.

It is clear that the different trucks in the unimodal transport example have similar performance regarding CO_2 emissions, but that there are much lower emissions of NO_x and PM with the Euro V truck. The multimodal chain dominated by train transport shows significantly lower emissions when green electricity is used. The diesel train gives high emissions of NO_X and PM. For electric trains, it is obvious that the source of electricity is essential for the outcome. The multimodal chain with a container ship and a barge results in lower CO_2 emissions but higher NO_X and PM emissions in comparison with the unimodal chain. The advantages with a larger ship and with Tier III for NO_X are obvious. The use of RoRo ships gives, in this example, somewhat lower emissions of CO_2 than the unimodal chain. The use of LNG results in lower emissions of NO_X and, especially, PM.

POLICY INSTRUMENTS

The factors deciding the mode of transport, and if a multimodal or a unimodal chain is chosen, are essentially the transport time, the accuracy of delivery and, of course, the price. For environmental factors to be more relevant, other pressures on the cargo owners and transporters are required. This can be in the form of legislation, economic incentives from authorities or initiatives from customers.

The EU white paper on transport (EU, 2011) strongly emphasises the need to move goods from road to rail and water, implying an increased use of intermodal transport. Thirty per cent of road freight over 300 km should shift to other modes such as rail or waterborne transport by 2030, and more than 50% by 2050. A major motivation for this is the problems with road congestion; reducing the emission of climate gases is also an important objective. Here, a goal is set to reduce the emissions of greenhouse gases by at least 60% by 2050 compared with 1990 levels. Multimodal transportation is one of the key strategies employed to solve the problems with the transport system in the EU. The white paper emphasises the use of the most efficient modes or combination of modes, and stipulates that transport users need to pay the full cost of the transport, that is, including externalities. Further, it is pointed out that existing infrastructure must be used more efficiently, which would lead to increased use of waterways and therefore intermodal transport. Also the previous white paper from 2001 pointed out the need to move goods from road to rail and waterways; however, the increase in road transportation has continued. Several policies are in place within the EU to promote intermodal transport. However, for environmental problems, they typically depend on intermodality itself to lead to improvements, rather than improving each mode. The so-called Eurovignette directive (Directive 2011/76/EU) allows EU member states to charge road traffic in order to cover infrastructure costs as well as societal costs from air pollution and noise. If this were to be used, it would promote the transfer from roads to other transport modes.

By far, emission legislation has been the most important policy instrument for the environmental impact from freight transportation. In the EU, the limits can be found in several regulations for emissions (Regulation 595/2009, 2012/46/EU) and fuel quality (2009/30/EC, 2012/33/EU). This legislation has led to major improvements when it comes to the emission of air pollutants from road traffic, and can be expected to have the same results for ships and diesel trains in the future. Policies driving the electrification of the railways have, of course, also been very significant.

Several policy measures can be used to improve the environmental performance of each mode as discussed, such as emission regulations, fuel standards, environmental zones, support for new technology and so on. One important way to increase the use of the most efficient mode from an environmental perspective is to internalise the external costs though taxes or fees. Macharis et al. (2010) conclude that this would lead to a competitive advantage for intermodal transportation.

CONCLUSION

Intermodal transport is often referred to as more environmentally friendly than unimodal transport. This is accurate if it means a transfer from road transport to other modes, such as rail and sea, especially for CO_2. For other emissions, modern trucks may actually show good performance when compared with the existing diesel train engines and ship engines. From an environmental perspective, there is no inherent advantage in using container transportation in relation to bulk transportation. The former essentially includes using fossil fuel, which is associated with emissions and other problems, in order to transport air. Therefore, the potential environmental advantage of intermodal systems lies in the use of common cargo carriers that facilitate the transfer from road to rail and sea. From the examples presented here, it is obvious that the potential environmental advantage of intermodal transportation in relation to unimodal transportation will depend on a number of details: the difference between old and new trucks is large when it comes to air pollutants; the size of the vessel is essential for fuel efficiency and CO_2 emissions; the type of electricity used for trains is essential, as is the fuel used for ships. Further, the load capacity utilisation will be important, and it is essential to keep this high throughout the multimodal chain. In addition, the advantages of a modal transfer from roads are obvious when it comes to congestion and noise.

So, what can be done to minimise the negative impact on the environment from intermodal transport systems? Starting with the emission of climate gases, all modes can minimise this by moving away from fossil fuels. As is obvious from Figure 13.1, rail systems using green electricity have radically reduced such emissions. Electricity may also be used for road transport in the near future and in limited applications for the sea mode (mainly for very short distances or as a replacement for auxiliary engines). Another option is to replace fossil fuels with biofuels; however, the effect in a life cycle perspective will depend on the methods for production of the fuel; for some fuels (e.g. ethanol produced from crops), the advantage may be small or non-existent (JEC, 2014). There are also a number of methods to reduce fuel consumption. For ships, the most important measure is to reduce the vessel speed; the fuel consumption per travelled distance is approximately proportional to the square of the speed. There are also a number of measures regarding design and technologies, but the fuel consumption per tonne-kilometre may also be reduced through measures to reduce empty trips and increase the load factors.

For noxious gases, there are a number of possibilities to reduce emissions: use of electricity, use of natural gas and applying exhaust gas after treatment methods. Trucks fulfilling the Euro VI standards must be considered as having low emissions due to extensive use of after treatment. Similar technologies as those used for trucks may also be applied to train engines and to ships. However, some of the technologies will put requirements on fuel quality that are currently not in place for marine fuels.

REFERENCES

European Commission. (2011). Roadmap to a single European transport area: Towards a competitive and resource efficient transport system. White Paper COM(2011) 144 final.

European Commission Joint Research Centre. (2014). Well-to-tank report version 4.0 JEC Well-to-Wheels Analysis. Luxembourg, Publications Office of the European Union.

Eurostat. (2016a). Table: Greenhouse gas emissions by source sector [env_air_gge]. Available at http://ec.europa.eu/eurostat/data/database

Eurostat. (2016b). Table: Air pollutants by source sector [env_air_emis]. Available at http://ec.europa.eu/eurostat/data/database

Harald M. Hjelle, Erik Fridell. (2012). When is short sea shipping environmentally competitive?, Environmental health - emerging issues and practice. In Prof. Jacques Oosthuizen (Ed.), InTech, DOI: 10.5772/38303. Available at http://www.intechopen.com/books/environmental-health-emerging-issues-and-practice/the-comparative-environmental-performance-of-short-sea-shipping

Kreutzberger, E., Macharis, C., Vereecken, L., Woxenius, J. (2003). Is intermodal freight transport more environmentally friendly than all-road freight transport? A review, NECTAR Conference No. 7, Umeå, Sweden.

Lowe, D. (2005). *Intermodal Freight Transport*. Amsterdam, Elsevier.

Macharis, C., Van Hoeck, E., Pekin, E., van Lier, T. (2010). A decision analysis framework for intermodal transport: Comparing fuel price increases and the internalisation of external costs. *Transportation Research Part A: Policy and Practice*. 44 (7): 550–561.

Moomaw, W., Burgherr, P., Heath, G., Lenzen, M., Nyboer, J., Verbruggen, A. (2011). Annex II: Methodology. In Edenhofer, O., Pichs-Madruga, R., Sokona, Y., Seyboth, K., Matschoss, P., Kadner, S., Zwickel, T., Eickemeier, P., Hansen, G., Schlömer, S., von Stechow, C. (eds.), IPCC Special Report on Renewable Energy Sources and Climate Change Mitigation. Cambridge, Cambridge University Press.

Stripple, H., Fridell, E., Winnes, H. (2016). Port infrastructures in a system perspective. IVL report C128.

Winebrake, J. J., Corbett, J. J., Falzarano, A., Scott Hawker, J., Korfmacher, K., Ketha, S., Zilora, S. (2008). Assessing energy, environmental, and economic tradeoffs in intermodal freight transportation. *Journal of the Air and Waste Management Association*. 58 (8): 1004–1013.

Index

Printed in the United States
by Baker & Taylor Publisher Services